This timely review provides a self-contained introduction to the mathematical theory of stationary black holes and a self consistent exposition of the corresponding uniqueness theorems.

The opening chapters examine the general properties of space-times admitting Killing fields and contain a detailed derivation of the Kerr Newman metric. Strong emphasis is given to the geometrical concepts. The general features of stationary black holes and the laws of black hole mechanics are then reviewed. Subsequently, critical steps towards the proof of the 'no hair' theorem are discussed, including the methods used by Israel, the divergence formulae derived by Carter, Robinson and others, and finally the sigma model identities and the positive mass theorem. The book is rounded off with an extension of the electrovacuum uniqueness theorem to self-gravitating scalar fields and harmonic mappings.

This volume provides a rigorous textbook for graduate students in physics and mathematics. It also offers an invaluable, up-to-date reference for researchers in mathematical physics, general relativity and astrophysics.

CAMBRIDGE LECTURE NOTES IN PHYSICS 6

General Editors: P. Goddard, J. Yeomans

Black Hole Uniqueness Theorems

CAMBRIDGE LECTURE NOTES IN PHYSICS

Black Hole Uniqueness Theorems

MARKUS HEUSLER
University of Zurich, Switzerland

CAMBRIDGE
UNIVERSITY PRESS

Published by the Press Syndicate of the University of Cambridge
The Pitt Building, Trumpington Street, Cambridge CB2 1RP
40 West 20th Street, New York, NY 10011–4211, USA
10 Stamford Road, Oakleigh, Melbourne 3166, Australia

© Cambridge University Press 1996

First published 1996

A catalogue record for this book is available from the British Library

Library of Congress cataloguing in publication data

Heusler, Markus.
Black hole uniqueness theorems / Markus Heusler.
p.   cm. – (Cambridge lecture notes in physics : 6)
Includes bibliographical references and index.
ISBN 0 521 56735 1 (pbk.)
1. Black holes (Astronomy) – Mathematics.   I. Title.   II. Series.
QB843.B55H48 1996
523.8'875'0151   dc20   96–11787 CIP

ISBN 0 521 56735 1 paperback

Transferred to digital printing 2003

Contents

Preface

In a manuscript communicated to the Royal Society by Henry Cavendish in 1783, an English scientist, Reverend John Michell, presented the idea of celestial bodies whose gravitational attraction was strong enough to prevent even light from escaping their surfaces. Both Michell and Laplace, who came up with the same concept in 1796, based their arguments on Newton's universal law of gravity and his corpuscular theory of light.

During the nineteenth century, a time when the notion of "dark stars" had fallen into oblivion, geometry experienced its fundamental revolution: Gauss and Lobachevsky had already found examples of non–Euclidean geometry, when Riemann became aware of the full consequences which arise from releasing the parallel axiom. In a famous lecture given at Göttingen University in 1854, the former student of Gauss introduced both the notion of spatial curvature and the extension of geometry to more than three dimensions.

It is these features of Riemannian geometry which, more than fifty years later, enabled Einstein to reveal the connection between the gravitational field and the metric structure of spacetime. In February 1916 - only three months after having achieved the final breakthrough in general relativity - Einstein presented, on behalf of Schwarzschild, the first exact solution of the new equations to the Prussian Academy of Sciences.

It took, however, almost half a century until the geometry of the Schwarzschild spacetime was correctly interpreted and its physical significance was fully appreciated. The neutron had yet to be discovered and the theory of stellar evolution to be developed such that neutron stars could be understood; only then would it become clear that there existed no physical laws to prevent certain stars from undergoing total gravitational collapse. This ultimate

fate of sufficiently massive stars had already been predicted in
the early 1930s by Chandrasekhar, who also elaborated the critical limit for the masses of white dwarf stars. At the present time
there is hardly any doubt concerning the existence of black holes
in the Universe. In fact, the picture given by Oppenheimer and
Snyder in 1939 has turned out to be in full agreement with current
knowledge:

> When all the thermonuclear sources of energy are exhausted a suf-
> ficiently heavy star will collapse. Unless fission due to rotation, the
> radiation of mass, or the blowing off of mass by radiation, reduce
> the star's mass to the order of that of the sun, this contraction will
> continue indefinitely... *Oppenheimer and Snyder (1939)*

The mathematical theory of black holes has been steadily de-
veloping during the last thirty years. One of its most intriguing
outcomes is the so-called "no-hair" theorem, which states that
a black hole in a stationary electrovacuum spacetime is uniquely
characterized by its mass, angular momentum and electric charge.
This result bears a striking resemblance to the fact that a statis-
tical system in thermal equilibrium is also described by a small
set of state variables, whereas considerably more information is
required to understand its dynamical behavior. This similarity is
reinforced by the black hole mass variation formula and the area
increase theorem, which are analogous to the corresponding laws
of ordinary thermodynamics. These mathematical relationships
are given physical significance from the observation that the temperature of the black-body spectrum of the Hawking radiation is
equal to the surface gravity of the black hole.

The purpose of this text is to provide an introduction to stationary black holes and to present a self-consistent exposition of
the corresponding uniqueness theorems. Although the emphasis
is given to the new approach to these theorems, based on the
positive energy theorem and sigma model identities, I have tried
to take the traditional line of reasoning into account as well. In
view of the recent developments in the field - notably the new
black hole solutions which reflect the limited realm of the classical
uniqueness theorems - some stress is laid upon the distinction between purely geometric results and conclusions which involve the
matter fields. The book starts out with some general properties

of spacetimes with Killing fields and with a systematic derivation of the Kerr–Newman metric. The body of the work is devoted to the properties of stationary black holes and their uniqueness theorems. The last chapters deal with self–gravitating mappings and include, in part, some recent results. I have also tried to provide a link with research in the field, by referring to problems which are currently under investigation.

This text is intended to be intelligible to the reader familiar with the basic notions of general relativity. Differential forms are used throughout, mainly for the sake of the improved efficiency of numerous derivations. It is therefore desirable that the reader is comfortable with both the calculus of tensor fields and differential forms. Most derivations are worked out in detail; however, I have given priority to a clear presentation of the geometrical concepts rather than to mathematical rigor.

I would like to acknowledge discussions with many colleagues. In particular, I wish to thank Robert Wald, Jürgen Ehlers and the Relativity Groups at the Enrico Fermi Institute in Chicago, the Max–Planck–Institute in Munich and the University of Zurich. I am very grateful to Piotr Chruściel for pointing out weak parts in the draft and providing me with the appropriate amendments. I owe especial thanks to Vivek Iyer for having read critically the manuscript and helping me to improve the style of this book with numerous valuable suggestions. Finally, I am particularly indebted to Norbert Straumann for many instructive discussions, his continuous support during the years and for drawing my attention to the uniqueness problem.

I gratefully acknowledge financial support from the Swiss National Science Foundation, the Max–Planck–Gesellschaft and the Tomalla Stiftung. It is a pleasure to thank Adam Black of the Cambridge University Press for his courtesy and cooperation.

I dedicate this book to my wife Regina, whose patience and understanding enabled me to write it.

1
Preliminaries

We assume that the reader is familiar with the fundamental notions of differential geometry. From a large number of mathematical texts we suggest the books of Helgason (1962), Bishop and Crittendon (1964), Kobayashi and Nomizu (1969), Matsushima (1972), Westenholz (1978), Spivak (1979), Choquet–Bruhat et al. (1982) and Willmore (1993) for comprehensive introductions into the subject. Concise accounts, designed to meet the needs of a relativist, can be found in Hawking and Ellis (1973), Kramer et al. (1980), Chandrasekhar (1983), Straumann (1984) and Wald (1984). The reader is referred to these books for introductions into the concepts of manifolds, tensor fields, connections and curvature. Since we intend to use an efficient formalism, we shall start this text with a brief review of the basic properties of differential forms. Before doing so, we fix some notations and conventions.

1.1 Conventions

Throughout this text, spacetime is denoted by $(M, \boldsymbol{g})$, where M is a 4–dimensional differentiable manifold endowed with a pseudo–Riemannian metric $\boldsymbol{g} = g_{\mu\nu} \, \theta^\mu \otimes \theta^\nu$ with signature $(-, +, +, +)$. (In order to avoid confusion, we occasionally write $^{(4)}\boldsymbol{g}$ instead of $\boldsymbol{g}$.) Greek indices label spacetime components of tensor fields, whereas Latin indices usually refer to components of lower–dimensional quantities.

The tangent space at a point $p \in M$ is denoted by $T_p(M)$ and the set of C^∞ vector fields X by $\mathcal{X}(M)$. We use the symbol ∇ for the unique metric, torsion–free affine connection on M, assigning the vector field $\nabla_X Y \in \mathcal{X}(M)$ to every pair of C^∞ vector fields $X, Y \in \mathcal{X}(M)$.

The curvature tensor maps the triple of fields $X, Y, Z \in \mathcal{X}(M)$

to the vector field $R(X,Y)Z \in \mathcal{X}(M)$. In terms of the connection ∇, the latter is defined by

$$R(X,Y)Z = \nabla_X(\nabla_Y Z) - \nabla_Y(\nabla_X Z) - \nabla_{[X,Y]}Z. \qquad (1.1)$$

The components of the Ricci tensor are obtained by the contraction

$$R_{\mu\nu} = R_{\mu\alpha\nu}{}^{\alpha}, \qquad (1.2)$$

where the components of the Riemann tensor are found from eq. (1.1). In terms of Christoffel symbols one has

$$R_{\mu\beta\nu}{}^{\alpha} = \partial_\beta \Gamma^{\alpha}_{\mu\nu} - \partial_\mu \Gamma^{\alpha}_{\beta\nu} + \Gamma^{\sigma}_{\mu\nu} \Gamma^{\alpha}_{\beta\sigma} - \Gamma^{\sigma}_{\beta\nu} \Gamma^{\alpha}_{\mu\sigma}. \qquad (1.3)$$

The commutation relations for the second covariant derivatives of a vector field $X \in \mathcal{X}(M)$ are

$$[\nabla_\nu \nabla_\mu - \nabla_\mu \nabla_\nu] X^{\beta} = R_{\mu\nu\alpha}{}^{\beta} X^{\alpha}. \qquad (1.4)$$

Let $\boldsymbol{T} = T_{\mu\nu}\theta^{\mu} \otimes \theta^{\nu}$ and $R = R_{\mu\nu}g^{\mu\nu}$ denote the stress–energy tensor of the matter fields and the Ricci scalar, respectively. The metric fields $g_{\mu\nu}$ are subject to Einstein's field equations

$$G_{\mu\nu} \equiv R_{\mu\nu} - \frac{1}{2} R g_{\mu\nu} = 8\pi G T_{\mu\nu}, \qquad (1.5)$$

which form a system of ten nonlinear, second order partial differential equations. The speed of light is set equal to 1 throughout this text.

1.2 Differential forms

Differential forms arise naturally in Riemannian geometry. Numerous formulae in general relativity are most efficiently obtained by deriving them within the framework of the exterior calculus. For instance, it is often easier to solve Cartan's structure equations than to compute the Riemann tensor from the Christoffel symbols. It is not our intention to provide an introduction into the exterior calculus in this section. Instead, we give only a brief account of the basic concepts and fix the conventions which we will need later. The reader who is not familiar with the subject should, for instance, consult Willmore (1993) for a concise introduction.

Consider an n–dimensional, orientable (pseudo–)Riemannian manifold $(M, \boldsymbol{g})$. Let $\Lambda(M) = \oplus_0^n \Lambda_p(M)$ denote the exterior algebra of differential forms on M. In a positively oriented local coordinate system $\{x^i\}$ $(i = 1...n)$ one has the representation

$$\alpha = \frac{1}{p!}\, \alpha_{\mu_1...\mu_p}\, dx^{\mu_1} \wedge ... \wedge dx^{\mu_p}\,, \tag{1.6}$$

$$\eta = \frac{1}{n!}\, \eta_{\mu_1...\mu_n}\, dx^{\mu_1} \wedge ... \wedge dx^{\mu_n} = \sqrt{|g|}\, dx^1 \wedge ... \wedge dx^n\,, \tag{1.7}$$

for an arbitrary p–form $\alpha \in \Lambda_p(M)$ and for the volume–form $\eta \in \Lambda_n(M)$, respectively. In terms of the determinant $g = \det(g_{\mu\nu})$ of the metric, the components of the volume–form are

$$\eta_{\mu_1...\mu_n} = \sqrt{|g|}\, \varepsilon_{\mu_1...\mu_n}\,, \tag{1.8}$$

where $\varepsilon_{\mu_1...\mu_n} = 1\,(-1)$ if $(\mu_1, ..., \mu_n)$ is an even (odd) permutation of $(1, ..., n)$, and $\varepsilon_{\mu_1...\mu_n} = 0$ otherwise.

Let us denote the interior multiplication of a p–form $\alpha \in \Lambda_p(M)$ by a vector field $X \in \mathcal{X}(M)$ by $i_X\alpha$, and let $d\alpha$ be the exterior derivative of α. The endomorphisms i_X and d are anti–derivatives $\Lambda(M) \to \Lambda(M)$ of degree -1 and 1, respectively:

$$(i_X\,\alpha)(X_1, ...X_{p-1}) = \alpha(X, X_1, ...X_{p-1}) \quad \in \Lambda_{p-1}(M)\,, \tag{1.9}$$

$$d\alpha = d\,(\alpha_{\mu_1...\mu_p}) \wedge dx^{\mu_1} \wedge ... \wedge dx^{\mu_p} \quad \in \Lambda_{p+1}(M)\,. \tag{1.10}$$

Note that the definition of $d\alpha$ is independent of the coordinate system. Also recall that $d\alpha = 0$ for $\alpha \in \Lambda_n(M)$ and $i_X\alpha = 0$ for $\alpha \in \Lambda_0(M)$. In addition, both i_X and d give zero when repeatedly applied to any p–form:

$$i_X \circ i_X\,\alpha = 0\,, \quad d \circ d\alpha = 0 \quad \text{for any } \alpha \in \Lambda_p(M)\,. \tag{1.11}$$

A p–form α is called *closed* if $d\alpha = 0$, and *exact* if there exists a $(p-1)$–form β such that $\alpha = d\beta$. Clearly every exact form is closed whereas, in general, the converse statement holds only locally. More precisely, Poincaré's lemma states that in a star–shaped domain every closed form is exact.

To every p–form α one can assign its Hodge dual, the $(n-p)$–form $*\alpha$, defined such that

$$\alpha \wedge \beta = (-1)^s\,(*\alpha\,|\,\beta)\,\eta\,, \tag{1.12}$$

for all $\beta \in \Lambda_{n-p}(M)$, where s denotes the number of negative eigenvalues of the metric. (Here we have used the natural extension of the inner product to each $\Lambda_p(M)$.) The components of $*\alpha$ are

$$(*\alpha)_{\mu_{p+1}...\mu_n} = \frac{1}{p!}\, \eta_{\mu_1...\mu_n}\, \alpha^{\mu_1...\mu_p}\,, \tag{1.13}$$

where $\alpha^{\mu_1 \cdots \mu_p} = g^{\mu_1 \nu_1} \cdot \ldots \cdot g^{\mu_p \nu_p} \alpha_{\nu_1 \ldots \nu_p}$. The inverse and the square of the Hodge dual of a p–form are given by

$$*^{-1} = (-1)^{p(n-p)+s} * , \qquad *^2 = (-1)^{p(n-p)+s} . \qquad (1.14)$$

Hence, for arbitrary p–forms α, $\beta \in \Lambda_p(M)$, one has the identity

$$(\alpha|\beta)\,\eta = \alpha \wedge *\beta = \beta \wedge *\alpha = (-1)^s \,(*\alpha\,|\,*\,\beta)\,\eta , \qquad (1.15)$$

since

$$*1 = \eta , \qquad *\eta = (-1)^s . \qquad (1.16)$$

To every p–form α one can also assign the $(p-1)$–form $d^\dagger \alpha \in \Lambda_{p-1}(M)$. In terms of the exterior derivative and the Hodge dual, the latter is defined by

$$d^\dagger \alpha = -(-1)^{n(p+1)+s} * d * \alpha . \qquad (1.17)$$

The component expression for $d^\dagger \alpha$ becomes

$$(d^\dagger \alpha)^{\mu_1 \cdots \mu_{p-1}} = -\frac{1}{\sqrt{|g|}} \partial_\mu \left(\sqrt{|g|} \, \alpha^{\mu \, \mu_1 \cdots \mu_{p-1}} \right) . \qquad (1.18)$$

The operator $d^\dagger : \Lambda_p(M) \to \Lambda_{p-1}(M)$ is called the co–derivative. The sign convention in the above definition is chosen such that d and $d^\dagger$ are formal adjoints of each other with respect to the inner product

$$\langle \, \cdot \, , \, \cdot \, \rangle = \int_M (\cdot\,|\,\cdot)\,\eta , \qquad (1.19)$$

that is, such that $\langle \, d\alpha \, , \, \beta \, \rangle = \langle \, \alpha \, , \, d^\dagger \beta \, \rangle$ for an orientable manifold M and $\alpha \in \Lambda_{p-1}(M)$, $\beta \in \Lambda_p(M)$.

A p–form α is said to be harmonic if both $d^\dagger \alpha$ and $d\alpha$ vanish and hence $\Delta \alpha = 0$, where, for arbitrary p–forms, the Laplacian is defined by

$$\Delta = - \left[d^\dagger d + d \, d^\dagger \right] . \qquad (1.20)$$

The sign convention in this definition is such that we obtain the usual coordinate expression

$$\Delta f = \frac{1}{\sqrt{|g|}} \partial_\nu (\sqrt{|g|} \, g^{\mu\nu} \, \partial_\mu f) \qquad (1.21)$$

for the Laplacian of a function f. (In the case of a pseudo–Riemannian metric the operator defined in eq. (1.20) is also called the d'Alembertian and is denoted by "$\Box$". Throughout this text we shall use the symbol Δ and, for any signature of the metric, refer to it as the Laplacian.)

Let us now restrict ourselves to the $3+1$–dimensional case. For $n = 4$ and $\mathrm{sig}(g) = (-,+,+,+)$ we find

$$*^{-1} = -(-1)^p *, \quad *^2 = -(-1)^p, \quad d^\dagger = *d* . \tag{1.22}$$

In particular, the co–derivative of the product of a function $f \in \Lambda_0(M)$ with a 1–form $\alpha \in \Lambda_1(M)$ becomes

$$d^\dagger(f\,\alpha) = f\,d^\dagger\alpha - (df|\alpha) . \tag{1.23}$$

In the following we shall often take advantage of the fact that the interior multiplication can be expressed in terms of the Hodge dual and the exterior product as

$$i_X\alpha = -*(X \wedge *\alpha), \quad i_X *\alpha = *(\alpha \wedge X), \tag{1.24}$$

where $\alpha \in \Lambda_p(M)$. Here we have used the symbol X for both the vector field $X \in \mathcal{X}(M)$ and the 1–form $X \in \Lambda_1(M)$ associated with it. As an application, we obtain the identity

$$\begin{aligned}(X|Y)\,(\alpha|\alpha) &= (X \wedge \alpha|Y \wedge \alpha) - (X \wedge *\alpha|Y \wedge *\alpha)\\ &= (i_X\alpha|i_Y\alpha) - (i_X *\alpha|i_Y *\alpha), \tag{1.25}\end{aligned}$$

which holds for arbitrary 1–forms $X, Y \in \Lambda_1(M)$ and arbitrary p–forms $\alpha \in \Lambda_p(M)$, since

$$\begin{aligned}(X \wedge \alpha|Y \wedge \alpha)\eta &= -i_Y *\alpha \wedge (X \wedge \alpha) = (-1)^p *\alpha \wedge i_Y(X \wedge \alpha)\\ &= (-1)^p\,[(X|Y)(*\alpha \wedge \alpha) - *\alpha \wedge X \wedge i_Y\alpha]\\ &= [(X|Y)\,(\alpha|\alpha) + (X \wedge *\alpha|Y \wedge *\alpha)]\,\eta .\end{aligned}$$

This establishes the first line in eq. (1.25). Applying eqs. (1.15) and (1.24), we also obtain the second part of the identity since $(X \wedge \alpha|Y \wedge \alpha) = -(i_X *\alpha|i_Y *\alpha)$ and $(X \wedge *\alpha|Y \wedge *\alpha) = -(i_X\alpha|i_Y\alpha)$. For $\alpha \in \Lambda_1(M)$, eq. (1.25) yields the familiar formula

$$(X \wedge \alpha|Y \wedge \alpha) = (X|Y)\,(\alpha|\alpha) - (X|\alpha)\,(Y|\alpha) . \tag{1.26}$$

2

Spacetimes admitting Killing fields

Einstein's field equations form a set of nonlinear, coupled partial differential equations. In spite of this, it is still sometimes possible to find exact solutions in a systematic way by considering space-times with symmetries. Since the laws of general relativity are covariant with respect to diffeomorphisms, the corresponding reduction of the field equations must be performed in a coordinate-independent way. This is achieved by using the concept of Killing vector fields. The existence of Killing fields reflects the symmetries of a spacetime in a coordinate-invariant manner.

A spacetime (M, g) admitting a Killing field gives rise to an invariantly defined 3-manifold Σ. However, Σ is only a hypersurface of (M, g) if it is orthogonal to the Killing trajectories. In general, Σ must be considered to be a quotient space M/G rather than a subspace of M. (Here G is the 1-dimensional group generated by the Killing field.) The projection formalism for M/G was developed by Geroch (1971, 1972a), based on earlier work by Ehlers (see also Kramer et $al.$ 1980). The invariant quantities which play a leading role are the twist and the norm of the Killing field.

In the first section of this chapter we compile some basic properties of Killing fields. The twist, the norm and the Ricci 1-form assigned to a Killing field are introduced in the second section. Using these quantities, we then give the complete set of reduction formulae for the Ricci tensor.

In the third section we apply these formulae to vacuum space-times. In particular, we introduce the vacuum Ernst potential and derive the entire set of field equations from a variational principle. As we shall see later, these equations reduce to the ordinary Ernst equations if spacetime admits two Killing fields satisfying the Frobenius integrability conditions.

The fourth and fifth sections are devoted to stationary and

static spacetimes, respectively. We recall the notions of Ricci–staticity and metric staticity and discuss the relationship between the two concepts. In the sixth section we derive some formulae for static spacetimes admitting a foliation by regular 2–surfaces. These will be relevant to later applications - especially to the original proof of the Israel theorem.

Spacetimes admitting two Killing fields are discussed in the last section of this chapter. In an asymptotically flat spacetime, the Killing fields generating the stationary and axisymmetric isometries form an Abelian group. This leads to the notion of stationary *and* axisymmetric spacetimes. We shall introduce the concepts of Ricci–circularity and metric circularity and discuss the integrability conditions for the Killing fields within these terms. In particular, we give a simple proof of the general circularity theorem, establishing the integrability conditions for Ricci–circular spacetimes. Some implications of the Frobenius conditions for the Killing 2–form assigned to a stationary and axisymmetric spacetime are discussed at the end of this chapter.

2.1 Killing fields

Consider the 1–parameter group of diffeomorphisms $\phi : M \to M$ generated by a vector field $X \in \mathcal{X}(M)$. A tensor field is invariant under ϕ if its Lie derivative with respect to X vanishes.

Definition 2.1 *A vector field $X \in \mathcal{X}(M)$ satisfying*

$$L_X \boldsymbol{g} = 0 \tag{2.1}$$

is called a Killing field.

We recall that the 1–parameter group of point transformations corresponding to a Killing field is an isometry.

In order to avoid an unnecessarily complicated notation, we shall use the same symbol for vector fields and their associated 1–forms. The following relations, which hold for arbitrary p–forms $\alpha \in \Lambda_p(M)$ and Killing fields (1–forms) K, turn out to be useful:

$$d^\dagger K = 0, \tag{2.2}$$

$$L_K * \alpha = *L_K \alpha, \tag{2.3}$$

$$d^\dagger(K \wedge \alpha) + K \wedge d^\dagger \alpha = -L_K \alpha. \tag{2.4}$$

The first equation is the contracted dual of the Killing equation for vector fields, $\nabla^\mu K^\nu + \nabla^\nu K^\mu = 0$. The commutation property of the Hodge dual with the Lie derivative with respect to a Killing field is easily verified for arbitrary p–forms. In order to derive the last relation, we apply the operator identity

$$L_K = d \circ i_K + i_K \circ d \qquad (2.5)$$

on the $(4-p)$–form $*\alpha$ and use eqs. (1.24) and (1.22):

$$\begin{aligned} L_K * \alpha &= -*(K \wedge *d * \alpha) + (-1)^p d * (K \wedge \alpha) \\ &= -*(K \wedge d^\dagger \alpha) - *d^\dagger (K \wedge \alpha). \end{aligned}$$

Taking the dual of this expression and using eq. (2.3) yields the desired result.

Definition 2.2 *The 1–form $R(X)$ assigned to a vector field $X \in \mathcal{X}(M)$ with components*

$$R(X)_\mu = R_{\mu\nu} X^\nu \qquad (2.6)$$

is called the Ricci 1–form with respect to X.

Contracting expression (1.4) and using the Killing equation, we see that the definition (1.20) of the Laplacian immediately yields the Ricci identity for a Killing field (1–form) K,

$$\Delta K = -2 R(K). \qquad (2.7)$$

We conclude this section by giving Stokes' theorem in the presence of a Killing field. The 1–dimensional group of isometries generated by the Killing field K give rise to an invariantly defined 3–manifold Σ (see Geroch 1971). Consider a 1–form α which is invariant under the action of this group, $L_K \alpha = 0$. Integrating the dual of the identity (2.4) and using Stokes' theorem ($\int d\cdot = \int_\partial \cdot$), we find for $\alpha \in \Lambda_1(M)$

$$\int_{\partial\Sigma} *(K \wedge \alpha) = -\int_\Sigma (d^\dagger \alpha)\, i_K \eta, \quad \text{if } L_K \alpha = 0, \qquad (2.8)$$

where $i_K \eta = *K$. This form of Stokes' theorem turns out be useful, for instance, in deriving integral identities relating the mass and the total charge of an electrovac spacetime (see chapter 8).

2.2 Basic identities

In this section we establish a set of differential identities between the Ricci 1–form, the twist and the norm of a Killing field. As we shall see later, these relations reduce to the Ernst equations for vacuum and electrovac spacetimes. In addition, they turn out to be the key identities for the proof of the staticity and circularity theorems.

Definition 2.3 *Let K be a Killing field (1–form). The function $N \in \Lambda_0(M)$ and the 1–form $\omega \in \Lambda_1(M)$,*

$$N = (K|K), \qquad \omega = \frac{1}{2} * (K \wedge dK) \qquad (2.9)$$

are called the norm and the twist (rotation–form) associated with K. In the domain of (M, g) where N is positive (negative, zero) the field K is said to be spacelike (timelike, null).

By virtue of eq. (1.24) and the above definition for ω, we have $-2 * (K \wedge \omega) = 2 i_K * \omega = i_K (K \wedge dK)$. Now using the fact that $i_K dK = -d i_K K = -dN$ (since $L_K K = 0$), we obtain the expression

$$- dK = \frac{1}{N} [2 * (K \wedge \omega) + K \wedge dN] \qquad (2.10)$$

for the derivative of the Killing form in terms of its twist and norm. Taking advantage of this formula, it is also not hard to verify the identity

$$N (dK|dK) = (dN|dN) - 4 (\omega|\omega). \qquad (2.11)$$

(First, the cross–terms do not contribute, since they are proportional to $K \wedge K$. Secondly, eq. (1.15) yields $(*[K \wedge \omega] | *[K \wedge \omega]) = -(K \wedge \omega|K \wedge \omega) = -N(\omega|\omega)$, since $(K|\omega) = 0$. Finally, eq. (1.26) implies $(K \wedge dN|K \wedge dN) = N(dN|dN)$, since $(K|dN) = L_K N = 0$.)

It is worthwhile pointing out that the Killing property of K has not been used so far. Hence, the above identities between the norm and twist hold for arbitrary vector fields (1–forms). However, in the remainder of this chapter, K is assumed to be a Killing field.

We now derive the formulae for the derivative and the co-derivative of the twist ω. We first note that both N and ω have vanishing Lie derivatives with respect to K, $L_K N = L_K \omega = 0$.

Hence, considering $\alpha = \omega/N^2$ in the general identity (2.4) and using the immediate consequence

$$- 2 * (K \wedge \frac{\omega}{N^2}) = d(\frac{K}{N})$$ (2.12)

of equation (2.10), we obtain

$$K \wedge d^\dagger (\frac{\omega}{N^2}) = -d^\dagger (K \wedge \frac{\omega}{N^2}) = \frac{1}{2} d^\dagger * d(\frac{K}{N}) = 0,$$

since $d^\dagger * d = \pm * d^2 = 0$. (Here we have used the fact that K is a Killing field, $d^\dagger K = 0$.) Thus, the norm and twist satisfy the differential identity

$$d^\dagger (\frac{\omega}{N^2}) = 0 \quad \text{or} \quad d^\dagger \omega = -2 \frac{(\omega|dN)}{N}$$ (2.13)

(i.e., $N\nabla^\mu \omega_\mu = +2\omega_\mu \nabla^\mu N$). Again using the identity (2.4) and $\Delta K = -d^\dagger dK = 0$, the exterior derivative of ω is obtained as follows:

$$2 \, d\omega = - * d^\dagger (K \wedge dK) = - * (K \wedge \Delta K).$$ (2.14)

In order to find an expression also involving the inner product of K with ΔK, we compute the Laplacian of N. By definition, we have $\Delta N = -d^\dagger dN$. Now using $dN = -i_K dK = *(K \wedge *dK)$ we obtain

$$\begin{aligned} - \Delta N &= * d (K \wedge *dK) = [(dK|dK) + (K|\Delta K)] * \eta \\ &= \frac{1}{N} [4 (\omega|\omega) - (dN|dN)] - (K|\Delta K), \end{aligned}$$ (2.15)

where we have also used $*\eta = -1$ and the relation (2.11) to substitute the quadratic term in dK. Taking advantage of the Ricci identity (2.7) enables us to replace the Laplacian of K in eqs. (2.14) and (2.15) by $-2R(K)$. In conclusion, we have established the following result:

Proposition 2.4 *The exterior derivative $d\omega$, the co–derivative $d^\dagger \omega$ and the square $(\omega|\omega)$ of the twist-form are given in terms of the Ricci 1–form $R(K)$ by*

$$d\omega = *(K \wedge R(K)),$$ (2.16)

$$d^\dagger \omega = -2 N^{-1}(\omega \,|\, dN),$$ (2.17)

$$(\omega|\omega) = \frac{1}{4} [(dN|dN) - N \Delta N - 2N R(K,K)].$$ (2.18)

It is worth noting that the integrability conditions for eq. (2.16) are fulfilled as a consequence of Bianchi's identity for the Einstein tensor and the Killing equation for K: By virtue of eq. (2.4) we have $d^2\omega = -*(K \wedge d^\dagger G(K)) = 0$, since the co–derivative of the Einstein 1–form $G(K)$ vanishes: $d^\dagger G(K) = -K^\nu \nabla^\mu G_{\mu\nu} - \frac{1}{2} G_{\mu\nu}(\nabla^\mu K^\nu + \nabla^\nu K^\mu) = 0$.

The above proposition yields the expressions for the $R(K, \cdot)$–components of the Ricci tensor. It remains to compute the expression for $R(X,Y)$, where X and Y are orthogonal to K. We do so by introducing the projection tensor P, defined in terms of the spacetime metric $^{(4)}g$ and the Killing field K by

$$^{(4)}g = \frac{1}{|N|} P + \frac{1}{N} K \otimes K. \qquad (2.19)$$

Note that P is positive definite if K is timelike, $|N| = -N$, whereas the signature of P is $(-, +, +)$ if K is a spacelike Killing field. The conformal factor in front of P turns out to be very convenient and is responsible for the relatively simple form of the Ricci tensor $R^{(P)}$ of M/G. The latter can be derived in an invariant manner or, of course, by writing the metric with respect to a coordinate basis adapted to $K = \partial_t$ (see, e.g., Israel and Wilson 1972 or Kramer *et al.* 1980). Since we shall present the reduction in some detail for the stationary and axisymmetric case, we give here only the result and invite the reader to verify it as an exercise. In terms of the Killing field K, its norm N and twist ω, one finds

$$R^{(P)} = R + \frac{1}{N} R(K,K) \, ^{(4)}g + \frac{1}{2N^2} \{dN \otimes dN + 4\omega \otimes \omega\}$$
$$- \frac{1}{N} \{K \otimes R(K) + R(K) \otimes K\}, \qquad (2.20)$$

where $^{(4)}g$ and R denote the metric and the Ricci tensor of spacetime. As a test, we observe that $L_K N = 0$ and $i_K \omega = 0$ imply $R^{(P)}(K, \cdot) = 0$. Hence, as required, $R^{(P)}$ vanishes unless it is evaluated on two vector fields orthogonal to K. Using Einstein's equations to express the Ricci tensor in terms of the stress–energy tensor, eqs. (2.16)-(2.20) yield the complete set of field equations for spacetimes admitting a Killing field.

2.3 The vacuum variational principle

In the previous section we demonstrated that both the component $R(K, K)$ of the Ricci tensor and the 2–form $K \wedge R(K)$ can be expressed in terms of the twist and the norm of the Killing field K. When $R(K, K)$ and $K \wedge R(K)$ are replaced by the corresponding stress–energy expressions, the identities (2.16)-(2.18) reduce to a set of differential equations - the Ernst equations - for N and ω on a 2–dimensional flat background manifold. This is, however, only the case if the spacetime admits *two* Killing fields which fulfil the integrability conditions. Nevertheless, the existence of *one* Killing field is sufficient to introduce the Ernst potential and to understand the basic structure of the Ernst equations. In addition, the entire set of vacuum (and electrovac) field equations for a spacetime with one Killing field can be obtained from a variational principle for the Ernst potential(s) and the projection metric $\boldsymbol{P}$.

We start by considering the complex 1–form $\mathcal{E}$, defined in terms of N and ω by

$$\mathcal{E} = -dN + 2i\,\omega. \tag{2.21}$$

The equation for the derivative of $\mathcal{E}$ is immediately obtained from the corresponding expression for $d\omega$. In order to find the co–derivative of $\mathcal{E}$, we add the expression (2.17) for $d^\dagger \omega$ to eq. (2.18) for $\Delta N = -d^\dagger dN$. This yields

$$d\mathcal{E} = 2i * (K \wedge R(K)), \tag{2.22}$$

$$d^\dagger \mathcal{E} - N^{-1}(\mathcal{E} \,|\, \mathcal{E}) = -2(K \,|\, R(K)). \tag{2.23}$$

If Einstein's equations (together with the invariance properties of the matter fields) imply that $*(K \wedge R(K))$ vanishes, then the first of the above equations implies the (local) existence of a potential E with $d\mathrm{E} = \mathcal{E}$. This is, of course, the case for vacuum models and, as we shall argue below, is also true for self–gravitating scalar fields.

For the remainder of this section we restrict ourselves to vacuum spacetimes. In this case, by virtue of eq. (2.22), we can introduce the vacuum Ernst potential $\mathrm{E} \in \Lambda_0(M)$, defined (up to an exact differential) by

$$d\mathrm{E} = \mathcal{E}. \tag{2.24}$$

In terms of E, eqs. (2.23) and (2.20) reduce to

$$\Delta E = 2 \frac{(dE|dE)}{E + \bar{E}},\tag{2.25}$$

$$R^{(P)} = 2 \frac{dE \otimes d\bar{E}}{E + \bar{E}},\tag{2.26}$$

with $N = -\mathrm{Re}(E)$. (There is no complex conjugation in the inner product in the first equation.) In a stationary *and* axisymmetric situation, the Laplacian and the scalar product only involve the conformally flat metric of the 2–dimensional manifold orthogonal to the Killing fields. As a consequence, the Ernst equation (2.25) then decouples from the remaining Einstein equations.

It is important to note that the above equations can be obtained from a variational principle. In particular, eq. (2.25) is obtained by varying the action $S[E, \bar{E}]$,

$$S[E, \bar{E}] = 4 \int_M \frac{(dE|d\bar{E})}{(E + \bar{E})^2} * 1,\tag{2.27}$$

with respect to the complex conjugate potential $\bar{E}$:

$$
\begin{aligned}
\delta_{\bar{E}} S[E, \bar{E}] &= \int \delta_{\bar{E}} \left[d\bar{E} \wedge * \frac{dE}{N^2} \right] \\
&= \int \left[d\delta\bar{E} \wedge * \frac{dE}{N^2} - 2 \, d\bar{E} \wedge * \frac{dE}{N^3} \frac{\partial N}{\partial \bar{E}} \delta\bar{E} \right] \\
&\doteq -\int \delta\bar{E} \left[d * \frac{dE}{N^2} - d\bar{E} \wedge * \frac{dE}{N^3} \right] \\
&= -\int \frac{\delta\bar{E}}{N^2} \left[\Delta E + \frac{(dE|dE)}{N} \right] * 1,
\end{aligned}
$$

where '$\doteq$' stands for equal up to boundary terms. Note that we have used $d * dE = \Delta E * 1$ and $\partial N/\partial \bar{E} = \partial N/\partial E = -\frac{1}{2}$. Hence, the vanishing of the variation with respect to $\bar{E}$ implies that the vacuum Ernst potential satisfies eq. (2.25).

The above action principle and the form of eq. (2.26) suggest that the effective Lagrangian for the entire set of vacuum equations is

$$
\begin{aligned}
L[E, \bar{E}, P] &= \sqrt{|P|} \left[R^{(P)} - 2 \frac{(dE|d\bar{E})^{(P)}}{(E + \bar{E})^2} \right] \\
&= \sqrt{|P|} \, P^{\mu\nu} \left[R^{(P)}_{\mu\nu} - 2 \frac{E_{,\mu} \bar{E}_{,\nu}}{(E + \bar{E})^2} \right].\tag{2.28}
\end{aligned}
$$

In fact, variations with respect to the projection metric $\boldsymbol{P}$ yield eq. (2.26) since, as usual, $\sqrt{|P|}P^{\mu\nu}\delta R^{(P)}_{\mu\nu}$ gives rise to a boundary term.

This concludes our brief introduction to the structure of the Ernst equations and the corresponding variational principle for vacuum spacetimes with one Killing field. As we shall see later, a completely analogous formulation exists for the Einstein–Maxwell system with a Killing field.

2.4 Asymptotic flatness and stationarity

The study of isolated systems in general relativity involves a detailed investigation of the properties of asymptotically flat spacetimes (Geroch 1970b, 1972b, 1976). The first major step towards an understanding of this concept was achieved by Bondi (1960), by examining the null hypersurfaces of spacetime (see also Bondi *et al.* 1962, Sachs 1962, 1964). Later on, Penrose (1963, 1964, 1965a) realized that the notion of infinity can be made precise by adding the future and past endpoints of null geodesics to spacetime. Considering a conformal transformation (with asymptotically vanishing conformal factor) then enables one to understand the asymptotic properties of the metric by studying the boundary of the conformal spacetime. Roughly speaking, a spacetime is said to be asymptotically flat if there exists a conformal transformation such that conformal infinity (spacelike and null) has similar properties to Minkowski spacetime.

For a detailed introduction into the asymptotic properties of the gravitational field we refer the reader to the literature. In particular, precise definitions of asymptotic flatness can be found in the review article of Newman and Tod (1980), the book of Wald (1984) and the work of Ashtekar and Hansen (1978) (see also Ashtekar 1980, Chruściel 1989a, 1989b). For our purposes it will, however, not be necessary to enter into the subtleties of the subject, since we restrict ourselves to stationary spacetimes. In this case, we avoid the whole issue of radiation, which allows us to adopt an intuitive notion of asymptotic flatness.

It is well known that the concept of local energy density in general relativity is vacuous. In an asymptotically flat spacetime it is, however, possible to obtain meaningful expressions for both the total 4–momentum and the total angular momentum tensor of an isolated system. Bartnik (1986) has given a rigorous treatment of the asymptotic decay conditions which guarantee that the ADM mass (Arnowitt *et al.* 1962),

$$M_{ADM} = \frac{1}{16\pi} \int_{S^2_\infty} (\partial_j\, g_{ij} - \partial_i\, g_{jj})\, dS^i, \qquad (2.29)$$

is well defined (see also Ashtekar 1984). Here $\partial_i \equiv \partial/\partial x^i$, where $\{x^i\}$ are asymptotically Euclidean coordinates on a spacelike hypersurface, and dS^i denotes the unit surface element of the 2–sphere at infinity, S^2_∞.

As already mentioned, we shall assume that spacetime is also stationary: For physical reasons, one expects that "sufficiently long" after the gravitational collapse of a star to a black hole, the latter settles down into a stationary configuration. A spacetime is said to be stationary if it admits a 1–parameter group of isometries with timelike orbits, that is, if it admits a timelike Killing field k. In order to distinguish situations with and without ergoregions, it is convenient to adopt the following definitions:

Definition 2.5 *An asymptotically flat spacetime is called stationary if it admits an asymptotically timelike Killing field k.*

Definition 2.6 *A domain of spacetime admitting a nowhere vanishing, timelike Killing field is called strictly stationary (or stationary in the strict sense).*

(Note that these notions are not used consistently in the literature: One also encounters the terms pseudo–stationary or stationary near infinity for stationary, and stationary for strictly stationary.) Note also that throughout this text an arbitrary Killing field is denoted by K, whereas the lower case letter k is reserved for the Killing field generating the stationary symmetry.

Beig and Simon (1980a, 1980b, 1981) have shown that an asymptotically flat *and* stationary metric is analytic at spacelike infinity. Their result was based on the assumption that - in an adapted asymptotically Cartesian coordinate system with $k = \partial/\partial x^0$ - the metric becomes flat like r^{-1} (or faster). Only recently were Kennefick and Ó Murchadha (1995) able to show that all asymptotically flat solutions which fall off slower than

r^{-1} involve gravitational radiation near spacelike infinity (see, e.g., Christodoulou and Ó Murchadha 1981). Hence, under rather weak decay and differentiability conditions, a stationary and asymptotically flat metric is analytic near infinity and, in terms of the mass M and the total angular momentum vector S^j, has the asymptotic expansion

$$g_{00} = -(1 - \frac{2\,M}{r}) + \mathcal{O}(r^{-2}), \tag{2.30}$$

$$g_{ij} = (1 + \mathcal{O}(r^{-1}))\,\delta_{ij} + \mathcal{O}(r^{-2}), \tag{2.31}$$

$$g_{0i} = -\epsilon_{ijk}\frac{4\,S^j}{r^3}\,x^k + \mathcal{O}(r^{-3}). \tag{2.32}$$

We shall also need the asymptotic form of the metric for stationary and axisymmetric spacetimes. Putting the z–axis along the symmetry axis, $S^1 = S^2 = 0$, $S^3 = J$, we obtain

$$g = -\left[1 - \frac{2M}{r} + \mathcal{O}(r^{-2})\right]dt^2 - \left[\frac{4J}{r}\sin^2\vartheta + \mathcal{O}(r^{-2})\right]dt d\varphi$$

$$+ \left[1 + \mathcal{O}(r^{-1})\right]\left(dr^2 + r^2\,(d\vartheta^2 + \sin^2\vartheta\,d\varphi^2)\right). \tag{2.33}$$

In a stationary spacetime, the total mass can also be expressed in terms of the Killing field k (Komar 1959, 1962):

Proposition 2.7 *Let (M, g) be an asymptotically flat, stationary spacetime with asymptotically timelike Killing field k, and let S_∞^2 denote the 2–sphere at spacelike infinity. Then the total mass M is given by*

$$M = -\frac{1}{8\pi G}\int_{S_\infty^2} *dk. \tag{2.34}$$

Proof In the above coordinate system one has $k = g_{0\mu}dx^\mu$, and the components of the metric do not depend on x^0. Thus, $*dk = \partial_i g_{00} * (dx^i \wedge dx^0)$ plus additional terms which do not contribute to the integral over S_∞^2. This yields

$$-\int_{S_\infty^2} *dk = \int_{S_\infty^2} \frac{\sqrt{|g|}}{2}\partial_i g_{00}\,[g^{i0}\,g^{j0} - g^{ij}\,g^{00}]\,\epsilon_{jkl}\,dx^k \wedge dx^l.$$

Using the asymptotic expansions (2.30)-(2.32) and taking only contributing terms into account yields

$$-\int_{S_\infty^2} *dk = \lim_{r\to\infty}\frac{G\,M}{r^3}\int_{S_\infty^2}\epsilon_{jkl}\,x^j\,dx^k \wedge dx^l = 8\pi\,G\,M,$$

where we have used $d[\epsilon_{jkl}x^j\,dx^k \wedge dx^l] = \epsilon_{jkl}dx^j \wedge dx^k \wedge dx^l = 6\,dx^1 \wedge dx^2 \wedge dx^3$ and Stokes' theorem in the last step. □

It will also turn out to be useful to convert the asymptotic Komar integral (2.34) into a volume integral. Let Σ denote a spacelike hypersurface extending from spacelike infinity to a bounding 2–surface $\mathcal{H}$. Using the Ricci identity for Killing fields (see eqs. (1.20) and (2.7)),

$$d * dk = - * \Delta k = 2 * R(k),$$

and Stokes' theorem, eq. (2.34) yields

$$M - M_H = - \frac{1}{4\pi G} \int_\Sigma *R(k), \qquad (2.35)$$

where M_H is the Komar expression evaluated over $\mathcal{H}$:

$$M_H = - \frac{1}{8\pi G} \int_\mathcal{H} *dk. \qquad (2.36)$$

This formula plays an important role in the derivation of the first law of black hole physics. If the inner boundary is the intersection of Σ with the Killing horizon, $H[k]$, generated by k, then one has $M_H = \frac{\kappa}{4\pi}\mathcal{A}$, where κ and $\mathcal{A}$ denote the surface gravity and the area of $\mathcal{H} = H[k] \cap \Sigma$, respectively (see chapter 6).

The above theorem establishes that the Komar mass (2.34) coincides with the mass parameter M appearing in the asymptotic expansion (2.30). This does, however, not prove that the ADM mass (2.29) - being defined under the sole requirement of asymptotic flatness - is equal to the Komar mass in the stationary case. A proof of this fact, based on an identity for the quadratic Landau–Lifshitz pseudotensor (Landau and Lifshitz 1971) and a conformal transformation, can be found in Beig (1978) (see also Ashtekar and Magnon–Ashtekar 1979).

2.5 Static spacetimes

Static spacetimes form an important class of stationary spacetimes. The uniqueness theorems for nonrotating black holes are, for instance, heavily based on staticity. In this section we discuss the relationship between the concepts of metric staticity and Ricci–staticity. We shall also derive the explicit expression for the Ricci tensor of a static metric, for use in later applications.

Definition 2.8 *A domain of a spacetime is called static if the Killing field k generating the stationary symmetry is hypersurface–orthogonal (metric staticity).*

In the static case there exists (locally) a 3–dimensional (spacelike) hypersurface Σ orthogonal to the trajectories of k. As a consequence of the Frobenius theorem (see, e.g., Matsushima 1972) one obtains the following important corollary:

Corollary 2.9 *A stationary domain is static if and only if $k \wedge dk = 0$, i.e., if the twist of the stationary Killing field vanishes.*

Proof The integrability condition $\omega = 0$ can also be obtained without using Frobenius' theorem: In section 2.2 we derived the identity (2.10) for dk in terms of its twist and norm. Using this relation, or the equivalent equation (2.12), we immediately find

$$\omega = 0 \quad \Longleftrightarrow \quad d(\frac{k}{N}) = 0. \tag{2.37}$$

Hence, provided that the twist of k vanishes, the 1–form k/N is closed and there exists (locally) a function $f \in \Lambda_0(M)$ such that $k = N df$. This implies that the vector k is orthogonal to the hypersurface $f = $ constant. $\qquad\square$

In the literature one also encounters the following, weaker notion of Ricci–staticity:

Definition 2.10 *A stationary spacetime is called Ricci–static if the Ricci 1–form $R(k)$ assigned to k is proportional to the Killing 1–form k, that is, if $R(k) \wedge k = 0$.*

In section 2.2 we established the identity (2.16),

$$d\omega = *(k \wedge R(k)),$$

between the Ricci 1–form and the derivative of the twist–form. This shows that metric staticity implies Ricci–staticity:

Corollary 2.11 *A static spacetime is Ricci–static.*

It is obvious that the converse statement requires a global investigation, since one has to show that $d\omega = 0$ implies $\omega = 0$. This problem was first investigated by Lichnerowicz (1955) and, in the context of black holes, was solved by Hawking and Ellis (1973) for vacuum spacetimes. We shall return to this question in section 8.2, where we present a simple proof of the fact that Ricci–staticity implies metric staticity. The argument is based on strict stationarity and also applies to spacetimes with nonconnected Killing horizons.

The remainder of this section is devoted to the derivation of the Ricci tensor of a static spacetime. In the domain where the hypersurface orthogonal Killing field k is timelike, the metric of a static spacetime assumes the form

$$^{(4)}g = - S^2 \, dt^2 + g, \tag{2.38}$$

where now g denotes the Riemannian metric on the 3–dimensional hypersurface Σ orthogonal to k. The Laplacian of an arbitrary function $f \in \Lambda_0(\Sigma)$ with respect to g is given by

$$\Delta^{(g)} f = \Delta f - S^{-1} (dS | df). \tag{2.39}$$

Using equation (2.18) with $N = -S^2$ and $\omega = 0$, we therefore have

$$S \, \Delta^{(g)} S = S \, \Delta S - (dS | dS)$$

$$= \frac{1}{2N} [(dN | dN) - N \Delta N] = R(k, k). \tag{2.40}$$

The remaining components of the Ricci tensor are most easily obtained by solving Cartan's structure equations,

$$d\theta^\mu + \omega^\mu{}_\nu \wedge \theta^\nu = 0, \tag{2.41}$$

$$\Omega^\mu{}_\nu = d\omega^\mu{}_\nu + \omega^\mu{}_\sigma \wedge \omega^\sigma{}_\nu, \tag{2.42}$$

in an orthonormal tetrad basis $\{\theta^\mu\}$ of $(M, {}^{(4)}g)$ ($\theta^0 = S \, dt$). Denoting the connection–forms of the 3–dimensional Riemannian manifold (Σ, g) with respect to θ^i by ${}^{(g)}\omega^i{}_j$, we find from the first structure equation (2.41)

$$\omega^0{}_i = (S^{-1} \nabla_i^{(g)} S) \, \theta^0, \qquad \omega^j{}_i = {}^{(g)}\omega^j{}_i, \tag{2.43}$$

where $\nabla^{(g)}$ denotes the covariant derivative with respect to g. Using this in equation (2.42) yields the curvature–forms

$$\Omega^0{}_i = (S^{-1} \nabla_j^{(g)} \nabla_i^{(g)} S) \, \theta^j \wedge \theta^0, \qquad \Omega^j{}_i = {}^{(g)}\Omega^j{}_i. \tag{2.44}$$

Using the relation $\Omega^\mu{}_\nu = \frac{1}{2} R^\mu{}_{\nu\sigma\rho} \theta^\sigma \wedge \theta^\rho$, the nonvanishing components of the Ricci tensor and the curvature scalar with respect to the static metric (2.38) become

$$R_{tt} = S\,\Delta^{(g)}S, \tag{2.45}$$

$$R_{ij} = R_{ij}^{(g)} - S^{-1}\,\nabla_j^{(g)}\nabla_i^{(g)}S, \tag{2.46}$$

$$R = R^{(g)} - 2\,S^{-1}\Delta^{(g)}S. \tag{2.47}$$

For later use, we also introduce the third–rank tensor R_{kij} on (Σ, g), defined by

$$R_{kij} = 2\,\nabla_{[i}^{(g)}R_{j]k}^{(g)} + \frac{1}{2}\,g_{k[i}\nabla_{j]}^{(g)}R^{(g)}. \tag{2.48}$$

We recall that in three dimensions the Weyl tensor vanishes, and the conformal properties of the 3–geometry are described by the Bach tensor R_{kij}. In fact, the following well–known theorem holds (see, e.g., Eisenhart 1949):

Theorem 2.12 *A 3–dimensional Riemannian manifold (Σ, g) is conformally flat if and only if $R_{kij} = 0$.*

Let us compute the components of R_{kij} in a vacuum spacetime, i.e., for the case where Einstein's equations imply that the 4–dimensional Ricci tensor vanishes. Using the identities (2.45)-(2.47) with $R_{\mu\nu} = 0$ and the expression (1.4) to eliminate the second covariant derivatives of $S_{|k} \equiv \nabla_k^{(g)}S$, we find

$$R_{kij} = \frac{2}{S^2}\,S_{|[j}S_{|i]|k} + \frac{1}{S}\,R_{jink}^{(g)}\,S^{|n}. \tag{2.49}$$

The second term can be simplified by taking advantage of the general formula

$$R_{jink}^{(g)} = 2\,g_{j[n}R_{k]i}^{(g)} - 2\,g_{i[n}R_{k]j}^{(g)} + R^{(g)}\,g_{j[k}g_{n]i} \tag{2.50}$$

for the Riemann tensor in three dimensions. Using eq. (2.46) again, we obtain the following expression for R_{kij} in terms of S and the covariant derivative with respect to g:

$$R_{kij} = \frac{4}{S^2}\,S_{|[j}S_{|i]|k} + \frac{2}{S^2}\,g_{k[i}S_{|j]|n}S^{|n}. \tag{2.51}$$

As we shall see in section 9.2, this expression turns out to be of considerable importance in establishing the relationship between conformal flatness of (Σ, g) and spherical symmetry of the 2–surfaces in Σ with constant S.

2.6 Foliations of static spacetimes

In the previous section we established the formulae (2.45)-(2.47), implying the identities

$$G_{tt} = \frac{1}{2} S^2 R^{(g)}, \qquad (2.52)$$

$$G_{ij} = G_{ij}^{(g)} + \frac{1}{S} \left(g_{ij} \Delta^{(g)} S - S_{|i|j} \right), \qquad (2.53)$$

$$G_{ti} = 0 \qquad (2.54)$$

for the Einstein tensor of a static spacetime with metric $^{(4)}g = -S^2 dt^2 + g$ (where $S_{|i|j} = \nabla_j^{(g)} \nabla_i^{(g)} S$). In view of Israel's proof of the uniqueness theorem for static black holes, we shall now derive some formulae for the foliation of the 3–dimensional space (Σ, g) by 2–dimensional surfaces $S = $ constant. We follow essentially the reasoning given by Israel (1967, 1968). (The reader may temporarily skip this section. The following formulae will only be needed in section 9.1, where the original proof of the static uniqueness theorem is explained.)

In the following, 2–dimensional quantities are characterized by a tilde and labeled by lower–case Latin indices from the beginning of the alphabet, $a, b \in \{2, 3\}$. Covariant derivatives with respect to the induced metric $\tilde{g}_{ab}$ are denoted by a semicolon.

Let us consider the strictly static domain where S is a well–defined, positive coordinate and the surfaces $S = $ constant are regular. Then the quantity ρ,

$$\rho = (dS|dS)^{-1/2}, \qquad (2.55)$$

vanishes nowhere. One can therefore introduce two functions x^2 and x^3 which are constant along the orthogonal trajectories to the 2–surfaces $S = $ constant in Σ. Choosing x^2 and x^3 to be the coordinates on $S = $ constant, the metric of (Σ, g) assumes the form

$$g = \rho^2 \, dS \otimes dS + \tilde{g}. \qquad (2.56)$$

Here both ρ and $\tilde{g} = \tilde{g}_{ab} dx^a dx^b$ depend on x^2, x^3 and S. It is now a straightforward exercise to compute the Christoffel symbols for the above metric. Introducing the extrinsic curvature K_{ab} of the embedded 2–dimensional surface $S = $ constant in Σ,

$$K_{ab} = \frac{1}{2} \rho^{-1} \partial_S \, \tilde{g}_{ab}, \qquad (2.57)$$

we obtain

$$\Gamma^S_{ab} = -\rho^{-1}K_{ab}, \quad \Gamma^a_{SS} = -\rho\,\tilde{g}^{ab}\rho_{,b}, \quad \Gamma^c_{ab} = \tilde{\Gamma}^c_{ab},$$
$$\Gamma^S_{SS} = \rho^{-1}\rho_{,S}, \quad \Gamma^S_{Sa} = \rho^{-1}\rho_{,a}, \quad \Gamma^a_{bS} = \rho\,K^a_b.$$

The Ricci tensor of $(\Sigma, \boldsymbol{g})$ is now obtained from the general expressions (1.2) and (1.3),

$$R^{(g)}_{SS} = -\rho\,(\,\tilde{\Delta}\rho + K_{,S} + \rho K_{ab}K^{ab}\,), \qquad (2.58)$$

$$R^{(g)}_{ab} = \tilde{R}_{ab} - \rho^{-1}\rho_{;ab} - K\,K_{ab} - \rho^{-1}\tilde{g}_{ac}K^c_{b,S}, \quad (2.59)$$

$$R^{(g)}_{aS} = \rho\,(\,K^b_{a;b} - K_{,a}\,), \qquad (2.60)$$

where $K = K^a_a$. In addition, eq. (2.53) also requires us to compute $\Delta^{(g)}S$ and the second covariant derivatives of S with respect to $\boldsymbol{g}$. Using eqs. (2.56) and the Γ–symbols given above, one finds

$$\Delta^{(g)}S = \rho^{-1}\,(\,K - \frac{\rho_{,S}}{\rho^2}\,) \qquad (2.61)$$

and

$$g_{SS}\,\Delta^{(g)}S - S_{|S|S} = \rho K\,,$$

$$g_{ab}\,\Delta^{(g)}S - S_{|a|b} = \frac{1}{\rho}\,(\,K\tilde{g}_{ab} - K_{ab} - \tilde{g}_{ab}\,\rho^{-2}\rho_{,S}\,),$$

$$g_{Sa}\,\Delta^{(g)}S - S_{|S|a} = \rho^{-1}\rho_{,a}\,. \qquad (2.62)$$

In order to compute the right hand sides of the basic equations (2.52)-(2.53), it remains to calculate the Einstein tensor and the Ricci scalar of $(\Sigma, \boldsymbol{g})$. Taking advantage of the above expressions, we have

$$G^{(g)}_{SS} = \frac{\rho^2}{2}\,(-\tilde{R} + K^2 - K_{ab}K^{ab}), \qquad (2.63)$$

$$G^{(g)}_{ab} = \tilde{G}_{ab} - K\,K_{ab} + \frac{1}{2}\tilde{g}_{ab}\,(\,K^2 + K_{ab}K^{ab}\,)$$

$$+ \frac{1}{\rho}\,(\tilde{g}_{ab}\tilde{\Delta}\rho - \rho_{;ab}) + \frac{1}{\rho}\,(\tilde{g}_{ab}K_{,S} - \tilde{g}_{ac}K^c_{b,S}\,), \quad (2.64)$$

$$G^{(g)}_{aS} = \rho\,(\,K^b_{a;b} - K_{,a}\,) \qquad (2.65)$$

and

$$R^{(g)} = \tilde{R} - (K^2 + K_{ab}K^{ab}) - \frac{2}{\rho}(\tilde{\Delta}\rho + K_{,S}\,). \qquad (2.66)$$

(We recall that eqs. (2.63) and (2.65) are also consequences of the Gauss and the Codazzi–Mainardi equations, respectively, describing the embedding of hypersurfaces.) We finally note that the Poisson equation (2.45) now reads

$$R_{tt} = S\Delta^{(g)}S = \frac{S}{\rho}\left(K - \frac{\rho_{,S}}{\rho^2}\right) = \frac{S}{\rho\sqrt{\bar{g}}}\partial_S\left(\frac{\sqrt{\bar{g}}}{\rho}\right). \qquad (2.67)$$

With respect to the static metric (2.38) and g according to eq. (2.56), the dual of the derivative of the Killing 1–form $k = -S^2dt$ becomes

$$* dk = -2S * (dS \wedge dt) = -2\rho^{-1}\tilde{\eta}, \qquad (2.68)$$

where $\tilde{\eta}$ is the volume–form on the 2–surfaces, $\tilde{\eta} = \sqrt{\bar{g}}dx^2 \wedge dx^3$. Hence, the Komar mass is obtained from the surface integral

$$M = \frac{1}{4\pi G}\int_{S^2_\infty}\rho^{-1}\tilde{\eta}. \qquad (2.69)$$

In section 9.1 we shall also integrate the above expression for $*dk$ over the 2–surface $S = 0$, which will show that $\rho^{-1}|_{S=0}$ is the surface gravity of the horizon of a static black hole.

As an application of the above formulae we compute the Laplacian of k. By virtue of eqs. (2.67) and (2.68) we have

$$d * dk = -2\,\partial_S\left(\frac{\sqrt{\bar{g}}}{\rho}\right)dS \wedge dx^2 \wedge dx^3 = -2S^{-1}\rho\,R_{tt}\,dS \wedge \tilde{\eta}.$$

Using the orthonormal tetrad fields $\theta^0 = Sdt$, $\theta^1 = \rho\,dS$, the Laplacian of k becomes ($\Delta k = - * d * dk$ since $d^\dagger k = 0$)

$$\Delta k = \frac{2}{S}R_{tt} * (\theta^1 \wedge \tilde{\eta}) = -2SR_{00}\theta^0 = -2R(k), \qquad (2.70)$$

which is in agreement with the Ricci identity (2.7) for Killing fields.

We conclude this section by computing $R_{\alpha\beta\gamma\delta}R^{\alpha\beta\gamma\delta}$. The fact that this invariant quantity must remain finite over the horizon is used in Israel's uniqueness proof (see section 9.1). Since $R^i{}_{0jk} = 0$, $R^i{}_{0j0} = S^{-1}S^{|i}{}_{|j}$ and $R^i{}_{jkl} = R^{(g)\,i}{}_{jkl}$, we have

$$\frac{1}{8}R_{\alpha\beta\gamma\delta}R^{\alpha\beta\gamma\delta} = \frac{1}{8}R^{(g)}_{ijkl}R^{(g)\,ijkl} + \frac{1}{2}S^{-2}S_{|ij}S^{|ij}. \qquad (2.71)$$

In three dimensions the Riemann and the Einstein tensor fulfil the identity $\frac{1}{8}R^{(g)}_{ijkl}R^{(g)\,ijkl} = \frac{1}{2}G^{(g)}_{ij}G^{(g)\,ij}$. The vacuum Einstein equations $\Delta^{(g)}S = 0$ and $G_{ij}^{(g)} = S^{-1}S_{|ij}$ now imply the relation

$$\frac{1}{8}R_{\alpha\beta\gamma\delta}R^{\alpha\beta\gamma\delta} = S^{-2}S_{|ij}S^{|ij}.$$

Using eqs. (2.62) with $\Delta^{(g)}S = 0$ finally yields the result

$$\frac{1}{8}R_{\alpha\beta\gamma\delta}R^{\alpha\beta\gamma\delta} = \frac{1}{S^2\rho^2}\left[K^2 + K_{ab}K^{ab} + 2\frac{(\tilde{\nabla}\rho|\tilde{\nabla}\rho)}{\rho^2}\right].\qquad(2.72)$$

2.7 Stationary and axisymmetric spacetimes

The problem of finding stationary and axisymmetric solutions to Einstein's equations arises if one is interested in equilibrium configurations of rotating sources. The first asymptotically flat solution exhibiting these symmetries was given by Kerr (1963).

Later on, Tomimatsu and Sato (1972, 1973) presented a further class of stationary and axisymmetric exact solutions. Although the Tomimatsu–Sato solutions are asymptotically flat, they do not approach spherically symmetric configurations in the limit of vanishing angular momentum, since their static counterparts are the Weyl solutions (Weyl 1917, 1919). More seriously, all solutions of the Tomimatsu–Sato class (with nonvanishing quadrupole moment, $\delta \neq 1$) exhibit naked singularities.

In addition to the solutions mentioned above, there exists a variety of other stationary and axisymmetric solutions to Einstein's vacuum equations (see Kramer *et al.* 1980). However, they are all flawed in being either not asymptotically flat or having naked singularities or, in the best case, degenerate horizons.

We start this section by introducing the notions of axial symmetry, Ricci–circularity and (metric) circularity. We then give a simple proof of the circularity theorem. The theorem guarantees that Ricci–circularity implies integrability of the 2–dimensional subspaces orthogonal to the Killing fields generating the stationary and axisymmetric isometries. The metric of a circular spacetime can be written in a standard form, involving no off–diagonal terms between the 2–dimensional subspaces (Papapetrou 1953). The remainder of this section is devoted to the derivation of some identities for the Killing 2–form.

According to definition 2.5, a spacetime is stationary if it admits an asymptotically timelike Killing field k. Let us now consider spacetimes which exhibit the following additional symmetry:

Definition 2.13 *A spacetime is called cyclic symmetric if it is invariant under the action π of the 1-parameter group $SO(2)$. A cyclic symmetric spacetime is called axisymmetric if the fixed point set of π (i.e., the rotation axis) is nonempty.*

Definition 2.14 *A spacetime is called stationary and axisymmetric if it is both stationary and axisymmetric, and if the Killing fields generating the symmetries commute with each other.*

Throughout this text, the Killing field (1-form) generating the cyclic symmetry will be denoted by m. A theorem due to Carter (1970) guarantees that the group generated by k and m is Abelian if spacetime is asymptotically flat and admits a symmetry axis:

Theorem 2.15 *An asymptotically flat, stationary, axisymmetric spacetime with Killing fields k and m is stationary and axisymmetric, that is, $[k, m] = 0$.*

Since we are dealing exclusively with asymptotically flat spacetimes in this text, it is understood that k and m commute when using the term *stationary and axisymmetric* in the following. (We also refer to Schmidt 1978, Xanthopoulos 1978 and Chruściel 1993 in this context.)

The total mass (2.34) of a stationary spacetime can be expressed in terms of a flux integral involving the stationary Killing field. Similarly, one obtains the total angular momentum, J, of an axisymmetric spacetime,

$$J = \frac{1}{16\pi G} \int_{S_\infty^2} *dm \,, \qquad (2.73)$$

where m denotes the Killing 1-form generating the axial symmetry. Note that the Komar expressions for M and J differ by a factor of 2 (and a sign). It is worth pointing out that this is not an accident, but has a deep significance: Wald (1993b) and Iyer and Wald (1994) have shown that the ADM mass of a diffeomorphism invariant theory is not equal to the Noether charge associated with asymptotic time translations. In contrast, the Noether charge corresponding to asymptotic rotations *does* coincide with the canonical angular momentum. (The different signs in the definitions of M and J reflect the Lorentz signature of the metric.) We also refer the reader to Chruściel (1987) for information on angular momentum at spacelike infinity.

Using Stokes' theorem and eqs. (1.20) and (2.7) for $K = m$, the above definition yields

$$J - J_H = \frac{1}{8\pi G} \int_\Sigma *R(m) \,, \tag{2.74}$$

where, as earlier, Σ denotes a 3–dimensional manifold extending from spacelike infinity to a 2–dimensional boundary $\mathcal{H}$, and J_H is defined by

$$J_H = \frac{1}{16\pi G} \int_\mathcal{H} *dm \,. \tag{2.75}$$

Stationary spacetimes with a hypersurface orthogonal Killing field are called static. The notion of circularity is introduced in a similar way:

Definition 2.16 *A stationary and axisymmetric spacetime is said to be circular if the 2–surfaces orthogonal to the Killing fields generating stationarity and axisymmetry are integrable, that is, if they are tangent to 2–dimensional surfaces.*

The integrability conditions are now expressed in terms of the twist 1–forms ω_k and ω_m assigned to the Killing fields k and m, respectively:

Corollary 2.17 *A stationary and axisymmetric spacetime is circular if and only if* $m^{[\mu}k^\nu \nabla^\sigma k^{\rho]} = k^{[\mu}m^\nu \nabla^\sigma m^{\rho]} = 0$, *that is, if*

$$(m|\omega_k) = (k|\omega_m) = 0 \,. \tag{2.76}$$

Proof Frobenius' theorem guarantees that the hypersurfaces orthogonal to k and m are integrable if and only if there exists a set of four 1–forms $\alpha_i \in \Lambda_1(M)$ with

$$dk = \alpha_1 \wedge k + \alpha_2 \wedge m \,, \quad dm = \alpha_3 \wedge k + \alpha_4 \wedge m \,,$$

i.e., if both $dk \wedge k \wedge m$ and $dm \wedge m \wedge k$ vanish (see, e.g., Bishop and Crittendon 1964, Matsushima 1972 or Wald 1984). This is, however, equivalent to the condition (2.76), since $(m|\omega_k)\eta = m \wedge *\omega_k = \frac{1}{2} m \wedge k \wedge dk$.   □

As in the static case, it turns out to be convenient to introduce the weaker notion of Ricci–circularity:

Definition 2.18 *A stationary and axisymmetric spacetime is said to be Ricci–circular if $m \wedge k \wedge R(k) = k \wedge m \wedge R(m) = 0$ or, equivalently, if*

$$i_m * (k \wedge R(k)) = i_k * (m \wedge R(m)) = 0. \qquad (2.77)$$

It may be helpful to compare the concepts of staticity ($\omega_k = 0$) and Ricci–staticity ($*(k \wedge R(k)) = 0$) on the one hand, and circularity and Ricci–circularity on the other hand: The circularity conditions (2.76) and (2.77) are obtained from the corresponding staticity conditions by a contraction with the second Killing field. In analogy with the static case (see corollary 2.11) we have the following corollary:

Corollary 2.19 *A circular spacetime is Ricci–circular.*

Proof The simplicity of the proof of this (and the converse) statement is an example of the efficiency of the calculus with forms: Clearly $[k, m] = 0$ implies $L_m \omega_k = L_k \omega_m = 0$. Using $L_m = i_m d + d i_m$ and the fundamental identity (2.16), $d\omega_k = *(k \wedge R(k))$, we immediately obtain

$$d(m|\omega_k) = d i_m \omega_k = -i_m d\omega_k = -i_m * (k \wedge R(k)), \qquad (2.78)$$

which demonstrates that $(m|\omega_k) = 0$ implies Ricci–circularity.   □

As mentioned above, a similar corollary for static spacetimes was established in section 2.5. However, there we had to postpone the proof of the converse statement, since it involves global arguments to show that the vanishing of the 2–form $d\omega_k$ implies the vanishing of the 1–form ω_k. Here the situation is much simpler since Ricci–circularity (2.77) is equivalent to the vanishing of a 1–form. In order to conclude from this that the integrability conditions (2.76) are satisfied, one has to establish the vanishing of a *function* (rather than a 1–form as in the static case). This is the reason why the following circularity theorem (Kundt and Trümper 1966,

Carter 1969; see also Carter 1987) is considerably simpler to prove than the corresponding staticity theorem (see section 8.2):

Theorem 2.20 *Let (M, g) be an asymptotically flat, stationary and axisymmetric spacetime. Then Ricci–circularity implies that the Killing fields fulfil the Frobenius integrability conditions (circularity) and vice versa.*

Proof From equation (2.78) we conclude that Ricci–circularity implies $d(m|\omega_k) = 0$. Hence, the function $(m|\omega_k) \in \Lambda_0(M)$ is constant. Since spacetime is asymptotically flat and axisymmetric, there exists a 2–dimensional set of points (the rotation axis) where m vanishes. Hence, $(m|\omega_k)$ vanishes identically (in every domain of (M, g) containing a part of the axis). Since the same argument applies also to the function $(k|\omega_m)$, we conclude that the 2–surfaces orthogonal to k and m are integrable, i.e., that spacetime is circular.   □

Finally, let us derive some identities for the Killing 2–form Ω and its norm $\mathcal{N}$, defined by

$$\Omega = k \wedge m , \qquad \mathcal{N} = -(\Omega|\Omega) . \qquad (2.79)$$

The reader who is not interested in these derivations may skip the remainder of this section. In what follows, the only formula which will be used is eq. (2.85), which holds if the Killing fields fulfil the integrability conditions.

To start, we write the integrability conditions (2.76) in terms of the Killing 2–form Ω:

Proposition 2.21 *The integrability conditions $(k|\omega_m) = (m|\omega_k) = 0$ are equivalent to*

$$* \Omega \wedge *d\Omega = 0 . \qquad (2.80)$$

In addition, they imply the relation

$$(d\mathcal{N}|d\mathcal{N}) + \mathcal{N} (d\Omega|d\Omega) = 0 . \qquad (2.81)$$

Proof The first part of the proposition is obtained as follows: Using $i_m * \Omega = i_k * \Omega = 0$ and the identity (1.24), we have

$$\begin{aligned}
*d\Omega \wedge *\Omega &= *[m \wedge dk - k \wedge dm] \wedge *\Omega = i_m(*dk \wedge *\Omega) \\
&\quad -i_k(*dm \wedge *\Omega) = -i_m(dk \wedge \Omega) + i_k(dm \wedge \Omega) \\
&= 2\,i_m(m \wedge *\omega_k) + 2\,i_k(k \wedge *\omega_m) \\
&= 2\,[(m|\omega_k) * m + (k|\omega_m) * k]\,,
\end{aligned}$$

which demonstrates that eqs. (2.76) and (2.80) are equivalent. In order to prove the second assertion, we first note that

$$d\mathcal{N} = -*(*d\Omega \wedge \Omega)\,. \tag{2.82}$$

This is obtained from $d\mathcal{N} = *d^\dagger(k \wedge m \wedge *\Omega)$ after repeated application of eq. (2.4) with $L_k\Omega = L_m\Omega = 0$. Using the general identity (1.25) for the 1–form $*d\Omega$ and the 2–form Ω now yields (with $(d\mathcal{N}|d\mathcal{N}) = -(*d\mathcal{N}|*d\mathcal{N})$),

$$\begin{aligned}
(d\mathcal{N}|d\mathcal{N}) &= -(*d\Omega \wedge \Omega|*d\Omega \wedge \Omega) = -(*d\Omega|*d\Omega)\,(\Omega|\Omega) \\
&\quad -(*d\Omega \wedge *\Omega|*d\Omega \wedge *\Omega) = -\mathcal{N}\,(d\Omega|d\Omega)\,,
\end{aligned}$$

where we have used the first part of the proposition in the last step. $\qquad\square$

Let us finally derive the expressions for $(\Omega|\Delta\Omega)$ and $\Delta(\Omega|\Omega)$, generalizing the corresponding identities $(K|\Delta K) = -2R(K,K)$ and $\Delta(K|K) = (K|\Delta K) + (dK|dK)$ for the Killing 1–form K (see eqs. (2.7) and (2.15)). Note that these identities hold independently of the integrability conditions.

Proposition 2.22 *Let k and m denote two commuting Killing fields, $[k,m] = 0$. Let $\Omega = k \wedge m$ and $X = (m|m)$, $W = (m|k)$, $V = -(k|k)$. Then*

$$(\Omega|\Delta\Omega) = 2\,[2W\,R(m,k) + V\,R(m,m) - X\,R(k,k)]\,, \tag{2.83}$$
$$\Delta(\Omega|\Omega) = (d\Omega|d\Omega) + (\Omega|\Delta\Omega)\,. \tag{2.84}$$

Proof As a trivial consequence of the general identity (2.4) and $L_k dm = dL_k m = d[k,m] = 0$, we have $\Delta\Omega = k \wedge \Delta m - m \wedge \Delta k$. Hence, we obtain the first identity from

$$(k \wedge m \,|\, k \wedge \Delta m - m \wedge \Delta k) =$$
$$(k|k)(m|\Delta m) + (m|m)(k|\Delta k) - (k|m)\,[(m|\Delta k) + (k|\Delta m)]\,,$$

using $(m|\Delta k) = -2(m|R(k)) = -2R(k,m)$. The second identity follows from eq. (2.82) for commuting Killing fields:

$$
\begin{aligned}
\Delta(\Omega|\Omega) &= d^\dagger d\mathcal{N} = - * d(\Omega \wedge *d\Omega) \\
&= - * (d\Omega \wedge *d\Omega) - *(\Omega \wedge *\Delta\Omega) \\
&= [(d\Omega|d\Omega) + (\Omega|\Delta\Omega)](- * \eta).
\end{aligned}
$$

$\square$

If the integrability conditions are fulfilled, we can use the expression (2.81) for $(d\Omega|d\Omega)$ in eq. (2.84). Substituting $(\Omega|\Delta\Omega)$ with the help of equation (2.83) finally yields the following result:

Corollary 2.23 *Let k and m be two commuting Killing fields satisfying the integrability conditions $k \wedge m \wedge dm = m \wedge k \wedge dk = 0$, and let $\mathcal{N} = -(k \wedge m|k \wedge m)$. Then*

$$
(k|k)\,R(m,m) - 2(m|k)\,R(m,k) + (m|m)\,R(k,k)
$$

$$
= \frac{1}{2}\left[\Delta\mathcal{N} - \frac{(d\mathcal{N}|d\mathcal{N})}{\mathcal{N}}\right]. \tag{2.85}
$$

3

Circular spacetimes

Einstein's equations simplify considerably in the presence of a second Killing field. Spacetimes with two Killing fields provide the framework for both the theory of colliding gravitational waves and the theory of rotating black holes (Chandrasekhar 1991). Although they deal with different physical objects, the theories are, in fact, closely related from a mathematical point of view. Whereas in the first case both Killing fields are spacelike, there exists an (asymptotically) timelike Killing field in the second situation, since the *equilibrium* configuration of an isolated system is assumed to be stationary. It should be noted that many stationary and axisymmetric solutions which have no physical relevance give rise to interesting counterparts in the theory of colliding waves. We refer the reader to Chandrasekhar (1989) for a comparison between corresponding solutions of the Ernst equations. In this chapter we discuss the properties of circular manifolds, that is, asymptotically flat spacetimes which admit a foliation by *integrable* 2–surfaces orthogonal to the asymptotically timelike Killing field k and the axial Killing field m.

In the first section we argue that the integrability conditions imply that locally $M = \Sigma \times \Gamma$ and $^{(4)}g = \sigma + g$. Here (Σ, σ) and (Γ, g) denote 2–dimensional manifolds where, in an adapted coordinate system, the metrics σ and g do not depend on the coordinates of Σ.

In the second section we discuss the properties of (Σ, σ), the pseudo–Riemannian manifold spanned by the orbits of the 2–dimensional Abelian group generated by the Killing fields. We derive an expression for the 2–dimensional Laplacian of $\sqrt{-\sigma}$ and give a differential identity for the 1–form $\mathcal{E} = -dX + 2i\omega$. Throughout this chapter, X and ω denote the norm and the twist of the *axial* Killing field m.

The third section deals with the properties of the manifold $(\Gamma, \boldsymbol{g})$ orthogonal to the Killing fields. Introducing the Riemannian metric $\boldsymbol{\gamma}$ which is conformally related to $\boldsymbol{g}$ by the norm of the axial Killing field, $\boldsymbol{\gamma} = X\boldsymbol{g}$, the Ricci tensor of $(\Gamma, \boldsymbol{\gamma})$ can be expressed in terms of $\sqrt{-\sigma}$ and the Ernst 1–form $\mathcal{E}$. We conclude this section with a summary of the basic identities.

The differential identities for the Ricci tensor in terms of $\rho \equiv \sqrt{-\sigma}$, $\mathcal{E}$ and the metric $\boldsymbol{\gamma}$ are further simplified if ρ is harmonic with respect to the 2–dimensional Riemannian metric $\boldsymbol{\gamma}$ (or $\boldsymbol{g}$). We shall see in the last section that this enables one to introduce Weyl coordinates and to write the metric in the Papapetrou form (Papapetrou 1953, 1966). As an application we consider vacuum spacetimes. We argue that Einstein's equations essentially reduce to a boundary value problem for one complex function (the Ernst potential E, dE $= \mathcal{E}$) in a flat 2–dimensional background metric. Having solved this equation, the remaining components of the metric are obtained by quadrature. In view of later applications we conclude this chapter by writing these equations in terms of prolate spheroidal coordinates, which turn out to be very convenient when deriving the Kerr and the Kerr–Newman metric.

3.1 The metric

We consider an asymptotically flat, stationary and axisymmetric spacetime $(M,{}^{(4)}\boldsymbol{g})$ which admits a foliation by 2–dimensional integrable surfaces orthogonal to the Killing fields k and m. The integrability conditions then imply that M is locally a product manifold,

$$M = \Sigma \times \Gamma, \tag{3.1}$$

where $(\Gamma, \boldsymbol{g})$ and (Σ, σ) are 2–dimensional manifolds with Riemannian metric $\boldsymbol{g}$ and pseudo–Riemannian metric σ, respectively. Parametrizing $\Sigma = \mathbb{R} \times \mathrm{SO}(2)$ with coordinates t and φ, the Killing vector fields become $k = \partial/\partial t$ and $m = \partial/\partial \varphi$. Introducing the coordinates $\{x^\mu\}$,

$$x^0 = t, \; x^1 = \varphi \in \Sigma; \quad x^2, x^3 \in \Gamma, \tag{3.2}$$

and choosing an adapted local basis of 1–forms,

$$\theta^a = dx^a, \; a = 0, 1; \quad \theta^i = dx^i, \; i = 2, 3, \tag{3.3}$$

the spacetime metric becomes

$$^{(4)}g = \sigma_{ab}\,\theta^a \otimes \theta^b + g_{ij}\,\theta^i \otimes \theta^j . \tag{3.4}$$

It is crucial that both 2–dimensional metrics, σ and g, depend only on the coordinates $\{x^i\}$ of Γ,

$$\sigma_{ab} = \sigma_{ab}(x^i), \quad g_{jk} = g_{jk}(x^i). \tag{3.5}$$

Note also that the co–derivative of an arbitrary stationary and axisymmetric 1–form, $\alpha = \alpha(x^i)$, and the Laplacian of a stationary and axisymmetric function, $f = f(x^i)$, become, respectively

$$d^\dagger \alpha = d^{\dagger(g)}\alpha - \rho^{-1}(d\rho|\alpha) = \rho^{-1}\,d^{\dagger(g)}(\rho\,\alpha), \tag{3.6}$$

$$\Delta f = \Delta^{(g)}f + \rho^{-1}(d\rho\,|\,df). \tag{3.7}$$

Here $\rho = \sqrt{-\sigma}$, and $d^{\dagger(g)}$ and $\Delta^{(g)}$ denote the co–derivative and the Laplacian with respect to g. (The difference in the signs occurs because $d^\dagger \alpha = -\rho^{-1}(\rho\alpha^i)_{,i}$ whereas $\Delta f = +\rho^{-1}(\rho f^{,i})_{,i}$; see eqs. (1.18), (1.21).) The determinant, $-\rho^2$, of the pseudo–Riemannian metric σ coincides with the norm, $-\mathcal{N}$, of the Killing 2–form Ω, introduced in eq. (2.79),

$$\rho^2 = -\det(\sigma) = -(k \wedge m | k \wedge m) = \mathcal{N} . \tag{3.8}$$

3.2 The orbit manifold

In this section we discuss some properties of the 2–dimensional pseudo–Riemannian orbit manifold (Σ, σ) with metric

$$\sigma = -V\,dt^2 + 2W\,dtd\varphi + X\,d\varphi^2 , \tag{3.9}$$

where

$$-V = (k|k), \quad W = (k|m), \quad X = (m|m). \tag{3.10}$$

First note the following: Both pairs $(-V, \omega_k)$ and (X, ω_m) satisfy the general identities (2.16)-(2.18) for the norm N and the twist ω of an arbitrary Killing field K. Hence, one has to select the Killing field with respect to which the Ernst equations will be formulated. As we shall see below, the appropriate choice - although not the traditional one - is to consider the Killing field m which generates the axial symmetry. There are two reasons for this: First, the norm, $-V$, of k has no fixed sign if spacetime is not strictly stationary. Thus, the system of differential equations formulated on the basis of the Killing field k turns out to be singular at the boundaries of "ergoregions" (which exist for rotating solutions). Secondly, if electromagnetic fields are also taken into account, the electrovac Ernst equations can still be derived from an action principle. However, only the Ernst equations based on the axial Killing field can be obtained from a *positive definite* Lagrangian. Definiteness of the effective Lagrangian is, in turn, a necessary condition in order to apply the uniqueness proof for rotating electrovac black holes (Mazur 1982, Bunting 1983). If we choose m as the fundamental Killing field, the metric (3.9) can be written in the form

$$\sigma = -\frac{\rho^2}{X} dt^2 + X (d\varphi + A\, dt)^2, \qquad (3.11)$$

where we have eliminated V and W in favor of the quantities $\rho = \sqrt{VX + W^2}$ and $A = W/X$. Instead of the function A, we can also consider the twist $\omega \equiv \omega_m$. The differential identities for ω and X were derived in section 2.2, whereas the basic identity for the Laplacian of ρ was given at the very end of the previous chapter (recall that $\rho^2 = \mathcal{N}$). It remains to write these identities in terms of the differential operators associated with the metric g and to establish the connection between the twist ω and the function A.

Let us start with the second problem and find the expression for the twist 1–form in terms of the functions A, ρ and X. Using the Killing 1–form $m = X d\varphi + W dt = X(d\varphi + A dt)$ and the definition $\omega = (1/2) * (m \wedge dm)$, we obtain

$$\omega = \frac{1}{2}X^2 * (dA \wedge dt \wedge d\varphi). \qquad (3.12)$$

This can be further simplified, since

$$*(dA \wedge dt \wedge d\varphi) = \frac{1}{\sqrt{-\sigma}} * (dA \wedge \eta^{(\sigma)}) = -\frac{1}{\rho} *^{(g)} dA,$$

where $\eta^{(\sigma)}$ and $*^{(g)}$ denote the volume 2–form on (Σ, σ) and the Hodge dual with respect to the 2–dimensional metric g, respectively. (Note that for $n = 2$ and $s = 0$, eqs. (1.14) and (1.17) yield $(*^{(g)})^{-1} = (-1)^p *^{(g)}$, $(*^{(g)})^2 = (-1)^p$ and $d^{\dagger(g)} = - *^{(g)} d *^{(g)}$.) The twist assigned to m is thus related to the derivative of A by

$$\omega = -\frac{X^2}{2\rho} *^{(g)} dA. \qquad (3.13)$$

The metric function A can therefore be obtained from ω by integrating the equation

$$dA = 2\rho \, *^{(g)} \left(\frac{\omega}{X^2}\right). \tag{3.14}$$

The integrability condition for this equation is verified by using the identity $d^\dagger(\omega/X^2) = 0$ (see eq. (2.13)) and the expression (3.6) for the co–derivative $d^{\dagger(g)}$ with respect to g:

$$d^2 A = 2d \, *^{(g)} \left[\rho\frac{\omega}{X^2}\right] = -2 \, *^{(g)} \, d^{\dagger(g)}[\rho\frac{\omega}{X^2}] = -2 \, *^{(g)} \, \rho d^\dagger \left(\frac{\omega}{X^2}\right) = 0 \,.$$

Below we shall also need the consequence

$$dA \otimes dA = \left(\frac{2\rho}{X^2}\right)^2 \left[(\omega|\omega)^{(g)} \, g - \omega \otimes \omega\right] \tag{3.15}$$

of eq. (3.14). This is established by using $(*^{(g)}\omega)_i = \eta_{ij}^{(g)} \omega^j$, where $(\cdot|\cdot)^{(g)}$ and $\eta^{(g)}$ denote the inner product and the volume form with respect to g.

In order to obtain the differential identities for the remaining metric coefficients X and ρ of σ, we rewrite eq. (2.23) for the 1–form $\mathcal{E} = -dX + 2i\omega$ and eq. (2.85) for $\mathcal{N}$ in terms of the 2–dimensional operators $\Delta^{(g)}$ and $d^{\dagger(g)}$. Using $\mathcal{N} = \rho^2$ and eq. (3.7) gives

$$\Delta\mathcal{N} - \frac{(d\mathcal{N}|d\mathcal{N})}{\mathcal{N}} = 2\rho\Delta^{(g)}\rho \,, \tag{3.16}$$

$$\mathrm{tr}_\sigma \boldsymbol{R} = \frac{1}{\rho^2} \left[2WR(m,k) - XR(k,k) + VR(m,m)\right], \tag{3.17}$$

where $\mathrm{tr}_\sigma \boldsymbol{R} = \sigma^{ab} R_{ab}$. Again taking advantage of eq. (3.6) for the co–derivative with respect to g, the basic identities (2.23) and (2.85) now become, respectively

$$\frac{1}{\rho} d^{\dagger(g)}(\rho\mathcal{E}) = \frac{(\mathcal{E}|\mathcal{E})}{X} - 2R(m,m) \,, \tag{3.18}$$

$$\frac{1}{\rho} \Delta^{(g)}\rho = -\mathrm{tr}_\sigma \boldsymbol{R} \,. \tag{3.19}$$

Note that the above equations are purely geometric identities for two Killing fields which are subject to the integrability conditions. They may, of course, also be derived by computing the components of the Ricci tensor for the metric (3.4), (3.11). As a simple exercise, we suggest the reader verifies the trace equation (3.19), using the Christoffel symbols

$$\Gamma^a_{bc} = \Gamma^a_{ij} = \Gamma^i_{aj} = 0\,,$$

$$\Gamma^a_{bi} = \tfrac{1}{2}\sigma^{ac}\,\partial_i\sigma_{bc}\,, \quad \Gamma^i_{ab} = -\tfrac{1}{2}g^{ij}\,\partial_j\sigma_{ab}\,. \tag{3.20}$$

3.3 The orthogonal manifold

In the previous section we expressed the components of the Ricci tensor of $(\Sigma, \boldsymbol{\sigma})$ in terms of the Ernst 1–form $\mathcal{E} = -dX + 2i\omega$, the quantity $\rho = \sqrt{-\sigma}$ and the metric $\boldsymbol{g}$. It remains to compute the components of the Ricci tensor of the 2–dimensional Riemannian manifold $(\Gamma, \boldsymbol{g})$ orthogonal to the Killing fields. This can be achieved by a further reduction of the general formula (2.20) with respect to the second Killing field. Alternatively, we can use the Christoffel symbols (3.20) to derive the relation between $R^{(g)}_{ij}$ and the Ricci tensor of spacetime. Since we did not give the derivation of the general formula (2.20), we present here the second method and verify the result by using eq. (2.20) at the end of the derivation.

In terms of the Christoffel symbols (3.20) derived from the metric (3.11), one finds (with $f_i \equiv \nabla^{(g)}_i f = \partial_i f$ for a stationary and axisymmetric function, $L_k f = L_m f = 0$):

$$
\begin{aligned}
R_{ij} &= R^{(g)}_{ij} - (\partial_i\Gamma^a_{aj} - \Gamma^a_{ak}\Gamma^k_{ij}) - \Gamma^a_{bi}\Gamma^b_{aj} \\
&= R^{(g)}_{ij} - \frac{1}{\rho}\nabla^{(g)}_i\nabla^{(g)}_j\rho + \frac{1}{\rho^2}\rho_i\rho_j + \frac{1}{4}\partial_i\sigma^{ab}\partial_j\sigma_{ab} \\
&= R^{(g)}_{ij} - \frac{1}{\rho}\nabla^{(g)}_i\nabla^{(g)}_j\rho + \frac{X_iV_j + V_iX_j + 2W_iW_j}{4\,\rho^2}.
\end{aligned}
\tag{3.21}
$$

In order to simplify this expression, it is convenient to introduce the metric $\boldsymbol{\gamma}$ on Γ, which is obtained from $\boldsymbol{g}$ by a conformal transformation with the norm of the axial Killing field,

$$\gamma_{ij} = X\,g_{ij}\,. \tag{3.22}$$

(Clearly $\boldsymbol{\gamma}$ is a positive Riemannian metric in the region of spacetime where m is spacelike.) The Ricci tensor $R^{(\gamma)}_{ij}$ and the second covariant derivative $\nabla^{(\gamma)}_i\nabla^{(\gamma)}_j f$ in terms of the corresponding quantities $R_{ij}^{(g)}$ and $\nabla_i^{(g)}\nabla_j^{(g)}f$ are (see, e.g., Wald 1984)

$$R_{ij}^{(g)} = R_{ij}^{(\gamma)} + \frac{1}{2}\gamma_{ij}\Delta^{(\gamma)}\ln X, \qquad (3.23)$$

$$\nabla_i^{(g)}\nabla_j^{(g)}f = \nabla_i^{(\gamma)}\nabla_j^{(\gamma)}f + \frac{1}{2X}[f_iX_j + f_jX_i - \gamma_{ij}(df|dX)^{(\gamma)}], \quad (3.24)$$

$$\Delta^{(g)}f = X\,\Delta^{(\gamma)}f, \qquad (3.25)$$

where $(\cdot|\cdot)^{(\gamma)} = X^{-1}(\cdot|\cdot)^{(g)}$ denotes the inner product with respect to γ. Using these expressions in eq. (3.21) one finds after some simple algebra

$$R_{ij} = \frac{1}{2}g_{ij}\left[\Delta^{(g)}\ln X + (d\ln\rho|d\ln X)^{(g)}\right] + \frac{X^2}{2\rho^2}A_iA_j$$
$$+ R_{ij}^{(\gamma)} - \frac{1}{\rho}\nabla_i^{(\gamma)}\nabla_j^{(\gamma)}\rho - \frac{1}{2X^2}X_iX_j. \qquad (3.26)$$

The term in brackets is equal to the 4–dimensional Laplacian of $\ln X$, for which we use the general expression (2.18). Substituting for the quadratic term in the derivatives of A by using eq. (3.15), the first line in the above equation becomes

$$-g_{ij}\left[\frac{R(m,m)}{X} + 2\frac{(\omega|\omega)^{(g)}}{X^2}\right] + 2\frac{(\omega|\omega)^{(g)}g_{ij} - \omega_i\omega_j}{X^2}$$
$$= -\frac{1}{X^2}[\gamma_{ij}R(m,m) + 2\omega_i\omega_j].$$

Using this in eq. (3.26) we obtain the desired result

$$R_{ij}^{(\gamma)} - \frac{1}{\rho}\nabla_i^{(\gamma)}\nabla_j^{(\gamma)}\rho = R_{ij} + \gamma_{ij}\frac{R(m,m)}{X^2} + \frac{X_iX_j + \omega_i\omega_j}{2\,X^2}, \quad (3.27)$$

where the last term can also be written as $(2X)^{-2}(\mathcal{E}_i\overline{\mathcal{E}}_j + \mathcal{E}_j\overline{\mathcal{E}}_i)$.

We can also derive this identity by evaluating the projection formula (2.20) for the axial Killing field m, also using the second symmetry in the projection metric P: First, eq. (2.19) with $K = m = X\,d\varphi + W\,dt$ and $N = X \geq 0$ yields the expression

$$P = -\rho^2 dt^2 + \gamma \qquad (3.28)$$

for the projection metric in terms of the 2–dimensional metric γ. (Here we have used $^{(4)}g = \sigma + \frac{1}{X}\gamma$ and eq. (3.9) for the metric of the orbit space.) Hence, the (i,j)–components of the Ricci tensor for the metric P become

$$R_{ij}^{(P)} = R_{ij}^{(\gamma)} - \frac{1}{\rho}\nabla_i^{(\gamma)}\nabla_j^{(\gamma)}\rho. \qquad (3.29)$$

On the other hand, these components can be read off from eq. (2.20): Using $K = m$, $^{(4)}g_{ij} = g_{ij} = X^{-1}\gamma_{ij}$ and $m_i = 0$, this gives

$$R_{ij}^{(P)} = R_{ij} + \gamma_{ij}\frac{R(m,m)}{X^2} + \frac{X_i X_j + \omega_i\omega_j}{2\,X^2}\,. \qquad (3.30)$$

Comparing the above two equations confirms the formula (3.27).

We conclude this section with a summary of the identities derived so far in this chapter:

Corollary 3.1 *Consider an asymptotically flat, stationary and axisymmetric spacetime $(M,^{(4)}g)$ admitting a foliation by integrable 2–surfaces orthogonal to the Killing fields. Then locally $M = \Sigma \times \Gamma$ and $^{(4)}g = \sigma + g$, and there exist coordinates t, φ such that $k = \partial/\partial t$, $m = \partial/\partial\varphi$ and*

$$^{(4)}g = -\frac{\rho^2}{X}\,dt^2 + X\,(d\varphi + A\,dt)^2 + \frac{1}{X}\gamma\,, \qquad (3.31)$$

where (Σ,σ) and (Γ,γ) are 2–dimensional manifolds with pseudo–Riemannian metric σ and Riemannian metric γ, respectively. The functions ρ, A and X are subject to the identities

$$\frac{1}{\rho}d^{\dagger(\gamma)}(\rho\mathcal{E}) = \frac{1}{X}\left[(\mathcal{E}|\mathcal{E})^{(\gamma)} - 2\,R(m,m)\right], \qquad (3.32)$$

$$\frac{1}{\rho}\Delta^{(\gamma)}\rho = -\frac{1}{X}\,\mathrm{tr}_\sigma\boldsymbol{R}\,, \qquad (3.33)$$

$$\frac{1}{\rho}dA = 2\,*^{(\gamma)}\left(\frac{\omega}{X^2}\right), \qquad (3.34)$$

$$\kappa^{(\gamma)}\gamma_{ij} - \frac{1}{\rho}\nabla_i^{(\gamma)}\nabla_j^{(\gamma)}\rho = R_{ij} + \gamma_{ij}\frac{R(m,m)}{X^2} + \frac{\mathcal{E}_i\bar{\mathcal{E}}_j + \mathcal{E}_j\bar{\mathcal{E}}_i}{4\,X^2}\,, \qquad (3.35)$$

where $d^{\dagger(\gamma)}$, $\Delta^{(\gamma)}$, $*^{(\gamma)}$ and $\kappa^{(\gamma)}$ denote the co–derivative, the Laplacian, the Hodge dual and the Gauss curvature with respect to the 2–dimensional metric γ. The complex 1–form $\mathcal{E}$ is defined by

$$\mathcal{E} = -\,dX + 2i\,\omega\,, \qquad (3.36)$$

and the twist ω associated to m fulfils the identity

$$d\omega = *(m \wedge R(m))\,. \qquad (3.37)$$

3.4 Weyl coordinates and the Papapetrou metric

Let us assume that the stationarity and axial symmetry of the matter fields imply that $\text{tr}_\sigma \boldsymbol{R}$ vanishes. Then, by virtue of eq. (3.33), $\rho = \sqrt{-\sigma}$ is a harmonic function on $(\Gamma, \boldsymbol{\gamma})$. In this case, the system given at the end of the previous section achieves a further, considerable simplification, which we discuss in this section. (We shall see later that $\text{tr}_\sigma \boldsymbol{R}$ vanishes for circular electromagnetic fields and for self–gravitating harmonic mappings. It does, however, not vanish for Yang–Mills fields with arbitrary gauge groups and for scalar fields with nonvanishing potentials.)

Our first goal is to show that ρ is a well–defined coordinate on $(\Gamma, \boldsymbol{\gamma})$. Using the Morse theory for harmonic functions (see, e.g., Milnor 1963), Carter (1973a, 1987) has argued that ρ has no critical points in the region where it is positive (i.e., in the domain of outer communications; see chapter 6). The following proof for this assertion was given by Weinstein (1990): For an asymptotically flat, simply connected domain of outer communications, the boundary of the region $\{0 < \rho < \rho_0\} \subset (\Gamma, \boldsymbol{\gamma})$ (with arbitrary positive ρ_0) consists of the two simple, nonclosed curves $\{\rho = 0\}$ and $\{\rho = \rho_0\}$. By virtue of the Riemann mapping theorem, there exists a *conformal* mapping, ϕ, from the above region into the strip $\{0 < \text{Re}(\zeta) < \rho_0\}$ in the complex ζ–plain. As a consequence of Caratheodory's extension of the mapping theorem (see, e.g., Behnke and Sommer 1976), we can also achieve that the boundaries are mapped into each other. Since in two dimensions $\Delta f = 0$ is invariant under conformal transformations, the pull–back of the harmonic function $f = \rho - \text{Re}(\zeta)$ to the ζ–plain by ϕ is again harmonic. Using the maximum principle and asymptotic flatness, one therefore finds $\rho = \text{Re}(\zeta)$, implying that ρ has no critical points in $\{\rho > 0\}$. As was recently shown by Chruściel and Wald (1994b), the domain of outer communications of a stationary black hole *is* indeed simply connected, provided that the null energy condition holds.

Hence, if $\Delta^{(\gamma)}\rho = 0$, it is possible to choose ρ as one of the coordinates on $(\Gamma, \boldsymbol{\gamma})$. Now considering the conjugate harmonic function z, one has $\Delta^{(\gamma)}z = 0$, $(d\rho|dz)^{(\gamma)} = 0$ and $(dz|dz)^{(\gamma)} = (d\rho|d\rho)^{(\gamma)}$. Hence, there exists a conformal factor $\exp(2h(\rho, z))$, such that the metric γ assumes the diagonal form

$$\gamma = e^{2h}(d\rho^2 + dz^2). \tag{3.38}$$

The spacetime metric (3.31) is now parametrized by the three functions X, A and h of the two variables ρ and z (Papapetrou 1953, 1966),

$$^{(4)}\boldsymbol{g} = -\frac{\rho^2}{X}dt^2 + X(d\varphi + A\,dt)^2 + \frac{1}{X}e^{2h}(d\rho^2 + dz^2). \tag{3.39}$$

It remains to rewrite eq. (3.35) in terms of the coordinates ρ and z. The second covariant derivatives of ρ transform into first partial derivatives of the conformal factor e^{2h} with respect to ρ and z. Using $\Gamma^\rho_{\rho\rho} = \Gamma^z_{z\rho} = -\Gamma^\rho_{zz} = \partial_\rho h$, $\Gamma^z_{zz} = \Gamma^z_{\rho z} = -\Gamma^z_{\rho\rho} = \partial_z h$, one immediately finds

$$\nabla^{(\gamma)}_z\nabla^{(\gamma)}_z\rho = -\nabla^{(\gamma)}_\rho\nabla^{(\gamma)}_\rho\rho = \partial_\rho h, \tag{3.40}$$

$$\nabla^{(\gamma)}_\rho\nabla^{(\gamma)}_z\rho = \nabla^{(\gamma)}_z\nabla^{(\gamma)}_\rho\rho = -\partial_z h, \tag{3.41}$$

and

$$\kappa^{(\gamma)} = -\Delta^{(\gamma)}h = -e^{-2h}\underline{\Delta}h. \tag{3.42}$$

Here $R^{(\gamma)}_{ij} = \kappa^{(\gamma)}\gamma_{ij}$, and $\underline{\Delta}$ denotes the flat Laplacian with respect to ρ and z. Hence, in terms of the metric (3.38), eq. (3.35) assumes the final form

$$\frac{1}{\rho}\partial_\rho h = \frac{1}{2}(R_{\rho\rho} - R_{zz}) + \frac{1}{4X^2}(\mathcal{E}_\rho\bar{\mathcal{E}}_\rho - \mathcal{E}_z\bar{\mathcal{E}}_z), \tag{3.43}$$

$$\frac{1}{\rho}\partial_z h = \frac{1}{2}(R_{\rho z} + R_{z\rho}) + \frac{1}{4X^2}(\mathcal{E}_\rho\bar{\mathcal{E}}_z + \mathcal{E}_z\bar{\mathcal{E}}_\rho), \tag{3.44}$$

$$-\underline{\Delta}h = \frac{1}{2}(R_{\rho\rho} + R_{zz}) + \frac{e^{2h}}{X^2}R(m,m) + \frac{1}{4X^2}(\mathcal{E}_\rho\bar{\mathcal{E}}_\rho + \mathcal{E}_z\bar{\mathcal{E}}_z). \tag{3.45}$$

The consistency and integrability conditions of this system are guaranteed by Bianchi's identities for the Ricci tensor and by the Ernst equation (3.32), which now reads

$$\partial_\rho\mathcal{E}_\rho + \partial_z\mathcal{E}_z + \frac{1}{\rho}\mathcal{E}_\rho + \frac{(\mathcal{E}|\mathcal{E})^{(\delta)}}{X} = 2\frac{1}{X}e^{2h}R(m,m). \tag{3.46}$$

As an example, we verify the integrability conditions for the vacuum case. Differentiating eq. (3.43) with respect to z and eq. (3.44) with respect to ρ yields ($R_{\rho\rho} = R_{zz} = R_{\rho z} = 0$)

$$\partial_{[z}\partial_{\rho]}h = -\rho\frac{\partial_z X}{2X^3}(\mathcal{E}_\rho\overline{\mathcal{E}}_\rho - \mathcal{E}_z\overline{\mathcal{E}}_z) - \frac{X - 2\rho\partial_\rho X}{4X^3}(\mathcal{E}_\rho\overline{\mathcal{E}}_z + \mathcal{E}_z\overline{\mathcal{E}}_\rho)$$

$$+\frac{\rho}{4X^2}[\partial_z(\mathcal{E}_\rho\overline{\mathcal{E}}_\rho - \mathcal{E}_z\overline{\mathcal{E}}_z) - \partial_\rho(\mathcal{E}_\rho\overline{\mathcal{E}}_z + \mathcal{E}_z\overline{\mathcal{E}}_\rho)]$$

$$= -\overline{\mathcal{E}}_z\frac{\rho}{4X^2}[\partial_\rho\mathcal{E}_\rho + \partial_z\mathcal{E}_z + \frac{1}{\rho}\mathcal{E}_\rho + \frac{\mathcal{E}_\rho^2 + \mathcal{E}_z^2}{X}] + \text{c.c.} = 0,$$

where c.c. denotes complex conjugation. In the last step we have used $\partial_i X = -\frac{1}{2}(\mathcal{E}_i + \overline{\mathcal{E}}_i)$ $(i = \rho, z)$ and the Ernst equation (3.46) with $R(m,m) = 0$. The consistency condition for eqs. (3.43)-(3.45), $\partial_\rho\partial_\rho h + \partial_z\partial_z h = \Delta h$, is established in a similar way.

This concludes our derivation of the basic identities for circular spacetimes. In the next chapter we shall impose the vacuum Einstein equations and derive the Kerr solution from the resulting system (3.32)-(3.37).

4

The Kerr metric

In the previous chapter we compiled the basic geometric identities for stationary and axisymmetric spacetimes. We shall now use these relations to derive the Kerr metric. Although we have to postpone the general definitions of black holes and event horizons to a later chapter, we feel that this is the right time to present the Kerr solution. As we shall argue when going into the details of the uniqueness theorems, the Kerr metric occupies a distinguished position amongst all stationary solutions of the vacuum Einstein equations.

The nonrotating counterpart of the Kerr solution was found by Schwarzschild (1916a, 1916b) immediately after Einstein's discovery of general relativity (Einstein 1915a, 1915b). In contrast to this, it took almost half a century until Kerr (1963) was eventually able to derive the first asymptotically flat exterior solution of a rotating source in general relativity. As is well known, both the Schwarzschild and the Kerr metric have charged generalizations, which were found by Reissner (1916) and Nordström (1918) in the static case, and by Newman *et al.* (1965) in the circular case.

The fact that it was not until 1963 that the Kerr metric was discovered reflects the difficulties of its derivation. As was pointed out by Chandrasekhar (1983), this does, however, not imply that "there is no constructive analytic derivation of the Kerr metric that is adequate in its physical ideas..." (Landau and Lifshitz 1971). In fact, the derivation of the Kerr solution appears fairly transparent when based on a discussion of the general properties of stationary and axisymmetric spacetimes.

We begin this chapter by stating the vacuum field equations for stationary and axisymmetric spacetimes. We then show that the Ernst equations can be obtained from a variational principle for the Gauss curvature of the orthogonal manifold (Γ, γ).

In the second section we introduce the notion of the conjugate solution $(\hat{X}, \hat{A}, \hat{h})$ and give the algebraic relations between the latter and a solution (X, A, h) of the vacuum equations.

Prolate spheroidal coordinates, $x = x(\rho, z)$ and $y = y(\rho, z)$, are introduced in the last section. In terms of x and y, the Ernst equations admit the simple, linear solution $\epsilon = px + iqy$, where $\epsilon = (1+\mathrm{E})/(1-\mathrm{E})$, and p and q are real parameters with $p^2 + q^2 = 1$. We show that the solution conjugate to $\epsilon = px + iqy$ yields the Kerr metric and conclude this chapter with a brief discussion of its asymptotic properties. This will also establish the relationship between the parameters p and q and the total mass M and angular momentum J.

4.1 The vacuum Ernst equations

In the previous chapter we derived the expressions for the Ricci tensor in terms of the Ernst 1–form $\mathcal{E} = -dX + 2i\omega$, the function $\rho = \sqrt{-\sigma}$ and the 2–dimensional metric γ (see eqs. (3.32)-(3.37)). As a first application, we shall now solve these equations for $R_{\mu\nu} = 0$, that is, we consider the vacuum Einstein equations

$$R_{\mu\nu} = 8\pi G \left[T_{\mu\nu} - \frac{1}{2}{}^{(4)}g_{\mu\nu} T \right] = 0. \qquad (4.1)$$

Using $R_{ij} = 0$, $R(m, m) = 0$ and $\mathrm{tr}_\sigma R = 0$ in corollary 3.1, we obtain the following proposition:

Proposition 4.1 *Let* $(M, {}^{(4)}g)$ *be an asymptotically flat, stationary and axisymmetric vacuum spacetime. Then* $(M, {}^{(4)}g)$ *is circular and the metric,* ${}^{(4)}g = \sigma + X^{-1}\gamma$, *can be parametrized in terms of three functions* X, A *and* h,

$$ {}^{(4)}g = -\frac{\rho^2}{X}dt^2 + X(d\varphi + A\,dt)^2 + \frac{1}{X}e^{2h}(d\rho^2 + dz^2). \qquad (4.2)$$

These functions are obtained from the complex Ernst potential E,

$$\mathrm{E}(\rho, z) = -X(\rho, z) + iY(\rho, z), \qquad (4.3)$$

by quadrature:

$$\frac{1}{\rho}\partial_\rho A = \frac{1}{X^2}\partial_z Y, \qquad \frac{1}{\rho}\partial_z A = -\frac{1}{X^2}\partial_\rho Y, \qquad (4.4)$$

$$\frac{1}{\rho}\partial_\rho h = \frac{1}{4X^2}[\partial_\rho E\partial_\rho\overline{E} - \partial_z E\partial_z\overline{E}],$$

$$\frac{1}{\rho}\partial_z h = \frac{1}{4X^2}[\partial_\rho E\partial_z\overline{E} + \partial_z E\partial_\rho\overline{E}]. \qquad (4.5)$$

The potential E *is subject to the Ernst equation,*

$$\frac{1}{\rho}\nabla(\rho\nabla E) + \frac{(\nabla E \mid \nabla E)^{(\delta)}}{X} = 0, \qquad (4.6)$$

which decouples from the remaining equations and is the Euler–Lagrange equation for the effective action

$$S_{\mathrm{vac}} = 4\int_\Gamma \kappa^{(\gamma)}\,\rho\,\eta^{(\gamma)} = 4\int_\Gamma \kappa^{(\gamma)}\sqrt{\gamma}\,\rho\,d\rho dz\,,$$

$$\text{with } \kappa^{(\gamma)}\sqrt{\gamma} = \frac{(\nabla E \mid \nabla\overline{E})^{(\delta)}}{(E + \overline{E})^2}, \qquad (4.7)$$

where $\kappa^{(\gamma)}$ *is the Gauss curvature of* (Γ,γ), $\delta = d\rho^2 + dz^2$, *and* $\nabla = (\partial_\rho, \partial_z)$.

Proof The Ricci–circularity conditions (2.77) are identically fulfilled in a vacuum spacetime. Hence, by virtue of the circularity theorem 2.20, the integrability conditions for the two Killing fields are satisfied as well, implying that M is circular with metric $^{(4)}g = \sigma + X^{-1}\gamma$. Since $\mathrm{tr}_\sigma R = 0$ implies that ρ is a harmonic function with respect to γ, one can introduce Weyl coordinates and write the metric in the Papapetrou form (3.39). The identity (3.37) immediately gives that the twist, ω, is closed. Hence, the Ernst 1–form, $\mathcal{E} = -dX + 2i\omega$, is closed as well, implying the local existence of a potential $E = -X + iY$ with $dE = \mathcal{E}$ and $dY = 2\omega$ (Ernst 1968a, 1968b). Eqs. (4.4) and (4.5) for A and h are now obtained from the corresponding identities (3.34), (3.43) and (3.44) with $R_{\mu\nu} = 0$. Finally, the vacuum Ernst equation (4.6) is obtained from eq. (3.46) or (3.32) with $\mathcal{E} = dE$ and

$$-\frac{1}{\rho}d^{\dagger(\gamma)}(\rho dE) = \Delta^{(\gamma)}E + (d\ln\rho|dE)^{(\gamma)} = \frac{1}{\rho}e^{-2h}\nabla(\rho\nabla E).$$

Since we have already established the integrability conditions for the metric functions A and h, it only remains to prove that the Ernst equation can be derived from the variational principle for the action (4.7). Using $(\,\cdot\,|\,\cdot\,)^{(\gamma)}\sqrt{\gamma}=(\,\cdot\,|\,\cdot\,)^{(\delta)}$, the action becomes

$$S_{\text{vac}} = \int_{\Gamma} \mathcal{L}_{\text{vac}}\, d\rho\, dz, \quad \mathcal{L}_{\text{vac}} = \rho\, \frac{(\nabla \text{E}\,|\,\nabla\overline{\text{E}})^{(\delta)}}{[\text{Re}(\text{E})]^2}. \tag{4.8}$$

It is not hard to verify that the Euler–Lagrange equation for $\mathcal{L}_{\text{vac}}$,

$$\left[\frac{\partial}{\partial\overline{\text{E}}} - \nabla\frac{\partial}{\partial(\nabla\overline{\text{E}})}\right]\mathcal{L}_{\text{vac}} = 0, \tag{4.9}$$

coincides with the Ernst equation (4.6). □

The last part of the above proof can also be obtained from the variational principle established in section 2.3 for the case of *one* Killing field: For $\text{E} = \text{E}(\rho, z)$, the action (2.27) with respect to the metric (4.2) reduces to

$$S[\text{E},\overline{\text{E}}] = \int_M 4\frac{(d\text{E}|d\overline{\text{E}})^{(\gamma)}}{(\text{E}+\overline{\text{E}})^2}\sqrt{|\sigma|}\sqrt{\gamma}\,dx^0\,dx^1\,d\rho\,dz$$

$$= \int_{\Gamma} \frac{(\nabla\text{E}|\nabla\overline{\text{E}})^{(\delta)}}{[\text{Re}(\text{E})]^2}\,\rho\,d\rho\,dz \int_{\Sigma} dx^0\,dx^1 = \text{vol}(\Sigma)\,S_{\text{vac}}.$$

In agreement with this, the general variational equation (2.25) for $S[\text{E},\overline{\text{E}}]$ reduces to the above vacuum Ernst equation when evaluated in the metric (4.2):

$$\Delta\text{E} - 2\frac{(d\text{E}|d\text{E})}{\text{E}+\overline{\text{E}}} = \Delta^{(g)}\text{E} + (d\text{E}|d\ln\rho)^{(g)} - 2\frac{(d\text{E}|d\text{E})^{(g)}}{\text{E}+\overline{\text{E}}}$$

$$= X\,e^{-2h}\,[\Delta\text{E} + \frac{1}{\rho}\partial_\rho\text{E} + \frac{1}{X}(\nabla\text{E}|\nabla\text{E})^{(\delta)}] = 0.$$

4.2 Conjugate solutions

In a moment we shall see that the Ernst equation (4.6) has a very simple solution, for which $\epsilon \equiv (1+\text{E})/(1-\text{E})$ is a linear function in terms of prolate spheroidal coordinates. The real part, $-X$, of the Ernst potential assigned to this solution does, however, not vanish on the rotation axis. Moreover, X is not positive semi–definite and has the wrong asymptotic behavior. It is, nevertheless, possible to obtain the Kerr metric from this solution by considering the so-called conjugate metric (see Chandrasekhar 1983). This additional step in the derivation of the Kerr solution becomes necessary if the Ernst equations are formulated with respect to the axial Killing field m. (As already mentioned, the Ernst system obtained on the basis of the stationary Killing field k exhibits other, more serious shortcomings.)

Proposition 4.2 *Consider a solution* (X, A, h) *of the vacuum field equations (4.4)-(4.6). Then the conjugate metric functions* $(\hat{X}, \hat{A}, \hat{h})$ *solve the same equations, where*

$$\hat{X} = X[A^2 - \frac{\rho^2}{X^2}], \quad \hat{A} = -A[A^2 - \frac{\rho^2}{X^2}]^{-1}, \quad e^{2\hat{h}} = e^{2h}\frac{\hat{X}}{X}. \quad (4.10)$$

Proof The basic observation consists of the fact that the metric σ of the orbit manifold is invariant under the (t, φ)-rotation

$$t \rightarrow \hat{t} = \varphi, \quad \varphi \rightarrow \hat{\varphi} = -t, \quad (4.11)$$

provided that X and A are transformed according to (4.10):

$$\begin{aligned}
\hat{\sigma} &= -\frac{\rho^2}{\hat{X}} d\hat{t}^2 + \hat{X}(d\hat{\varphi} + \hat{A}d\hat{t})^2 \\
&= \frac{\rho^2 X}{\rho^2 - X^2 A^2} d\varphi^2 - \frac{\rho^2 - X^2 A^2}{X}(dt - \frac{A X^2}{\rho^2 - X^2 A^2} d\varphi)^2 \\
&= -\frac{\rho^2}{X} dt^2 + X(d\varphi + Adt)^2 = \sigma.
\end{aligned}$$

It is clear that the transformation (4.11) does not affect the metric of the orthogonal manifold $(\Gamma, \boldsymbol{g})$. Hence, $\hat{\boldsymbol{g}} = \boldsymbol{g} = X^{-1}e^{2h}(d\rho^2 + dz^2)$ and thus $\hat{X}^{-1}e^{2\hat{h}} = X^{-1}e^{2h}$. $\square$

It is instructive to consider the following derivation of the transformation law for $e^{2\hat{h}}$: Let us assume that we are given a solution E of the vacuum Ernst equation (4.6), i.e., a pair (X, A) (see eqs. (4.3) and (4.4)). Transforming (X, A) according to the above formulae, we obtain a new solution $(\hat{X}, \hat{A})$, from which we compute $d\hat{Y}$ $(= 2\hat{\omega})$ by using eq. (4.4). This provides us with the conjugate Ernst 1-form $\hat{\mathcal{E}} = -d\hat{X} + 2i\hat{\omega}$. Now writing eq. (4.5) in terms of the conjugate quantities yields $\hat{h}$ by integration. In order to illustrate this procedure, we first derive the expression for the conjugate Ernst 1-form in terms of the unconjugated quantities. Differentiating the transformation laws for X and A, we have

$$\hat{\mathcal{E}} = -d\hat{X} - \frac{i}{\rho}\hat{X}^2 *^{(\gamma)} d\hat{A},$$

and thus

$$
\hat{\mathcal{E}} = 2\frac{\rho}{X}d\rho - (\frac{\rho^2}{X^2} + A^2)\left[dX + \frac{i}{\rho}X^2 *^{(\gamma)}dA\right]
$$
$$
+ 2iA *^{(\gamma)}d\rho - 2i\rho\frac{A}{X}\left[*^{(\gamma)}dX - \frac{i}{\rho}X^2 dA\right].
$$

Hence

$$
\hat{\mathcal{E}} = (\frac{\rho^2}{X^2} + A^2)\mathcal{E} + 2i\rho\frac{A}{X} *^{(\gamma)}\mathcal{E} + 2\frac{\rho}{X}d\rho + 2iAdz. \qquad (4.12)
$$

Using $*^{(\gamma)}\mathcal{E} = \mathcal{E}_\rho dz - \mathcal{E}_z d\rho$, we find, after some short algebraic manipulations,

$$
\frac{\hat{\mathcal{E}}_\rho\hat{\bar{\mathcal{E}}}_\rho - \hat{\mathcal{E}}_z\hat{\bar{\mathcal{E}}}_z}{\hat{X}^2} = \frac{\mathcal{E}_\rho\bar{\mathcal{E}}_\rho - \mathcal{E}_z\bar{\mathcal{E}}_z}{X^2} - \frac{2\rho}{X\hat{X}}\left[\frac{\bar{\mathcal{E}}_\rho + \mathcal{E}_\rho}{X} + iA\frac{\bar{\mathcal{E}}_z - \mathcal{E}_z}{\rho} + \frac{2}{\rho}\right]
$$

and

$$
\frac{\hat{\mathcal{E}}_\rho\hat{\bar{\mathcal{E}}}_z + \hat{\mathcal{E}}_z\hat{\bar{\mathcal{E}}}_\rho}{\hat{X}^2} = \frac{\mathcal{E}_\rho\bar{\mathcal{E}}_z + \mathcal{E}_z\bar{\mathcal{E}}_\rho}{X^2} - \frac{2\rho}{X\hat{X}}\left[\frac{\bar{\mathcal{E}}_z + \mathcal{E}_z}{X} - iA\frac{\bar{\mathcal{E}}_\rho - \mathcal{E}_\rho}{\rho}\right].
$$

Since $\bar{\mathcal{E}}_\rho + \mathcal{E}_\rho = -2\partial_\rho X$ and $i(\bar{\mathcal{E}}_\rho - \mathcal{E}_\rho) = 4\omega_\rho = -2\frac{X^4}{\rho^2}\partial_z A$, eq. (4.5) yields the following decoupled expressions for the partial derivatives of $\hat{h}$ in terms of h, X and A:

$$
\frac{1}{\rho}\partial_\rho\hat{h} = \frac{1}{\rho}\partial_\rho h - \frac{(X^2/\rho)AA_\rho + (\rho/X)X_\rho - 1}{\rho^2 - X^2A^2},
$$
$$
\frac{1}{\rho}\partial_z\hat{h} = \frac{1}{\rho}\partial_z h - \frac{(X^2/\rho)AA_z + (\rho/X)X_z}{\rho^2 - X^2A^2}.
$$

We thus obtain

$$
d\hat{h} = dh + \frac{1}{2}d[\ln(A^2 - \frac{\rho^2}{X^2})],
$$

which gives the desired result: $e^{2\hat{h}} = e^{2h}[A^2 - \frac{\rho^2}{X^2}] = e^{2h}X^{-1}\hat{X}$.

In the domain where X is negative, e^{2h} must become negative as well, since the signature of $g = X^{-1}e^{2h}(d\rho^2 + dz^2)$ is $(+,+)$. It is therefore convenient to use $\exp(2h)/X = \exp(2\hat{h})/\hat{X}$, and to write the metric of the orthogonal manifold in terms of conjugate quantities as well. The spacetime metric then assumes the form

$$
^{(4)}g = -\frac{\rho^2}{\hat{X}}d\hat{t}^2 + \hat{X}(d\hat{\varphi} + \hat{A}d\hat{t})^2 + \frac{e^{2\hat{h}}}{\hat{X}}(d\rho^2 + dz^2). \qquad (4.13)
$$

4.3 The Kerr solution

In this section we perform the last step in the derivation of the Kerr metric. We start by rewriting the vacuum Einstein equations (4.4)-(4.6) in terms of prolate spheroidal coordinates. As we shall see later, the region "outside" an axisymmetric black hole (the domain of outer communications) is characterized by the local requirement that the Killing 2-form, $\Omega = k \wedge m$, (see section 2.7) is timelike, i.e., that $\rho^2 = -(\Omega|\Omega)$ is positive; the exception being the rotation axis, where the Killing field m vanishes and hence $\rho = 0$. Since ρ also vanishes on the horizon, it is convenient to introduce prolate spheroidal coordinates x and y (see, e.g., Carter 1973a), defined in terms of ρ and z by

$$\rho^2 = \mu^2 (x^2 - 1)(1 - y^2), \qquad z = \mu x y, \qquad (4.14)$$

where μ is an arbitrary positive constant. Note that $(\rho, z) \mapsto (x, y)$ maps the upper half plane to the semi-strip $\{(x, y)| x \geq 1, |y| \leq 1\}$. The boundary $\rho = 0$ consists of the horizon ($x = 0$) and the northern ($y = 1$) and southern ($y = -1$) segments of the rotation axis. The Riemannian metric γ now becomes

$$\gamma = e^{2h}(d\rho^2 + dz^2)$$

$$= e^{2h} \mu^2 (x^2 - y^2) \left(\frac{dx^2}{x^2 - 1} + \frac{dy^2}{1 - y^2}\right). \qquad (4.15)$$

It is a straightforward task to rewrite the vacuum Ernst equation (4.6) (or (3.32)), as well as eq. (4.4) for A and eq. (4.5) for h, in terms of the new coordinates. One finds:

$$[(\rho\sqrt{\gamma}\gamma^{ij}) E_{,j}]_{,i} + \frac{1}{X} (\rho\sqrt{\gamma}\gamma^{ij}) E_{,i} E_{,j} = 0, \qquad (4.16)$$

$$A_{,x} = \mu (1 - y^2) \frac{Y_{,y}}{X^2}, \qquad A_{,y} = -\mu (x^2 - 1) \frac{Y_{,x}}{X^2}, \qquad (4.17)$$

$$h_{,x} = \frac{1 - y^2}{4X^2(x^2 - y^2)} [-y(x^2 - 1)(E_{,x}\overline{E}_{,y} + E_{,y}\overline{E}_{,x})$$
$$+ x(x^2 - 1) E_{,x}\overline{E}_{,x} - x(1 - y^2) E_{,y}\overline{E}_{,y}],$$

$$h_{,y} = \frac{x^2 - 1}{4X^2(x^2 - y^2)} [+x(1 - y^2)(E_{,x}\overline{E}_{,y} + E_{,y}\overline{E}_{,x})$$
$$+ y(x^2 - 1) E_{,x}\overline{E}_{,x} - y(1 - y^2) E_{,y}\overline{E}_{,y}], \qquad (4.18)$$

where $X = -\text{Re}(E)$, $Y = \text{Im}(E)$. It is an important fact that eq. (4.16) is an equation for the Ernst potential alone. In particular, as a consequence of the conformal invariance of $\sqrt{\gamma}\gamma^{ij}$ in two dimensions, it does not involve the function h.

In order to solve the above equations, it is convenient to introduce the potential ϵ, which maps the semi–plane $X = -\mathrm{Re}(E) \geq 0$ into the unit disc $|\epsilon| \leq 1$,

$$\epsilon \equiv \frac{1 + E}{1 - E}. \tag{4.19}$$

With respect to prolate spheroidal coordinates, the vacuum Einstein equations for ϵ, A and h finally assume the following form:

Corollary 4.3 *With respect to the metric*

$$
\begin{aligned}
^{(4)}g \;=\; & -\frac{\rho^2}{X}\, dt^2 \;+\; X\,(d\varphi + A dt)^2 \\
& + \frac{e^{2h}}{X}\,\mu^2\,(x^2 - y^2)\left(\frac{dx^2}{x^2 - 1} + \frac{dy^2}{1 - y^2}\right) \tag{4.20}
\end{aligned}
$$

of a circular spacetime, the vacuum field equations reduce to the vacuum Ernst equation,

$$
\begin{aligned}
&[(x^2 - 1)\,\epsilon_{,x}]_{,x} + [(1 - y^2)\,\epsilon_{,y}]_{,y} \\
&= -2\,\bar{\epsilon}\,(1 - \epsilon\bar{\epsilon})^{-1}\,[(x^2 - 1)\,\epsilon_{,x}^2 + (1 - y^2)\,\epsilon_{,y}^2], \tag{4.21}
\end{aligned}
$$

and the equations

$$A_{,x} = \mu\,(1 - y^2)\,\frac{Y_{,y}}{X^2}, \qquad A_{,y} = -\mu\,(x^2 - 1)\,\frac{Y_{,x}}{X^2}, \tag{4.22}$$

$$
\begin{aligned}
\frac{(1 - \epsilon\bar{\epsilon})^2(x^2 - y^2)}{(1 - y^2)}\,h_{,x} \;=\; & x\,[(x^2 - 1)\epsilon_{,x}\,\bar{\epsilon}_{,x} - (1 - y^2)\epsilon_{,y}\,\bar{\epsilon}_{,y}] \\
& - y\,(x^2 - 1)\,[\epsilon_{,x}\,\bar{\epsilon}_{,y} + \epsilon_{,y}\,\bar{\epsilon}_{,x}], \\
\frac{(1 - \epsilon\bar{\epsilon})^2(x^2 - y^2)}{(x^2 - 1)}\,h_{,y} \;=\; & y\,[(x^2 - 1)\epsilon_{,x}\,\bar{\epsilon}_{,x} - (1 - y^2)\epsilon_{,y}\,\bar{\epsilon}_{,y}] \\
& + x\,(1 - y^2)\,[\epsilon_{,x}\,\bar{\epsilon}_{,y} + \epsilon_{,y}\,\bar{\epsilon}_{,x}] \tag{4.23}
\end{aligned}
$$

for A and h, where the potentials X and Y are given in terms of ϵ by

$$X = \frac{1 - \epsilon\bar{\epsilon}}{|1 + \epsilon|^2}, \qquad Y = \frac{i\,(\bar{\epsilon} - \epsilon)}{|1 + \epsilon|^2}. \tag{4.24}$$

Eq. (4.21) can also be obtained from the variational principle for the effective action

$$S_{\text{vac}} = 4\mu \int_\Gamma \frac{(x^2 - 1)\,\epsilon_{,x}\,\bar{\epsilon}_{,x} + (1 - y^2)\epsilon_{,y}\,\bar{\epsilon}_{,y}}{(1 - \epsilon\bar{\epsilon})^2}\,dx\,dy\,. \qquad (4.25)$$

Proof The corollary is a direct consequence of proposition 4.1 and the preceding equations: The differential equations (4.23) for h are immediately obtained from eqs. (4.18) and

$$\frac{E_{,i}\,\overline{E}_{,j}}{4\,X^2} = \frac{\epsilon_{,i}\,\bar{\epsilon}_{,j}}{(1 - \epsilon\bar{\epsilon})^2}\,.$$

The Ernst equation (4.21) is derived from eq. (4.16). It can also be obtained from the variational principle for the effective action (4.7), which, in terms of ϵ and prolate spheroidal coordinates, becomes

$$\begin{aligned} S_{\text{vac}} &= 4 \int_\Gamma \frac{(dE|d\overline{E})^{(\gamma)}}{4\,X^2}\,\rho\,\eta^{(\gamma)} = 4 \int_\Gamma \frac{\epsilon_{,i}\,\bar{\epsilon}_{,j}\,(\gamma^{ij}\,\sqrt{\gamma}\rho)}{(1 - \epsilon\bar{\epsilon})^2}\,dx\,dy \\ &= 4\mu \int_\Gamma \frac{(x^2 - 1)\,\epsilon_{,x}\,\bar{\epsilon}_{,x} + (1 - y^2)\epsilon_{,y}\,\bar{\epsilon}_{,y}}{(1 - \epsilon\bar{\epsilon})^2}\,dx\,dy\,. \end{aligned}$$

$\qquad\qquad\qquad\qquad\qquad\qquad\qquad\qquad\qquad\qquad\qquad\qquad\square$

The Kerr solution is now derived as follows: First, the Ernst equation (4.21) admits the linear solution

$$\epsilon = px + i\,qy\,, \qquad (4.26)$$

where the real constants p and q are subject to the condition

$$p^2 + q^2 = 1\,. \qquad (4.27)$$

(Note that $\pm(px + iqy)^{-1}$ and $\pm(px - iqy)^{-1}$ are solutions of the vacuum equation (4.21) as well, which is most easily seen from the invariance of the effective action (4.25) with respect to inversion and complex conjugation of ϵ; see also section 5.3.) Substituting the solution (4.26) into the expressions (4.24) for X and Y gives

$$X = \frac{1 - (px)^2 - (qy)^2}{(1 + px)^2 + (qy)^2}\,, \qquad Y = \frac{2\,qy}{(1 + px)^2 + (qy)^2}\,, \qquad (4.28)$$

which yields

$$\begin{aligned} \partial_x A &= 2q\mu\,(1 - y^2)\frac{(1 + px)^2 - (qy)^2}{[1 - (px)^2 - (qy)^2]^2}\,, \\ \partial_y A &= 2p\mu\,(x^2 - 1)\frac{(1 + px)\,2qy}{[1 - (px)^2 - (qy)^2]^2}\,, \qquad (4.29) \end{aligned}$$

$$\partial_x h = -\frac{x\,(1-y^2)}{(x^2-y^2)\,[1-(px)^2-(qy)^2]}\,,$$

$$\partial_y h = -\frac{y\,(x^2-1)}{(x^2-y^2)\,[1-(px)^2-(qy)^2]}. \qquad (4.30)$$

The solutions are

$$A = 2\mu\,(q/p)\,\frac{(1-y^2)\,(1+px)}{1-(px)^2-(qy)^2} \qquad (4.31)$$

and

$$e^{2h} = \frac{1-(px)^2-(qy)^2}{p^2\,(x^2-y^2)}. \qquad (4.32)$$

As is seen from eq. (4.28), the norm, X, of the Killing field m does not vanish on the rotation axis. In addition, it is not positive definite on the whole semi–strip $\{|y| \le 1,\ x > 1\}$. (Note, however, that $X^{-1}e^{2h}$ is non–negative.) In order to obtain a physically acceptable solution, one must consider the conjugate quantities $\hat{X}$, $\hat{A}$ and $\hat{h}$. Performing the transformations (4.10) and using the algebraic identity

$$\frac{4q^2(1-y^2)(1+px)^2 - p^2(x^2-1)\,[(1+px)^2+(qy)^2]^2}{1-(px)^2-(qy)^2}$$

$$= [(1+px)^2 + q^2]^2 - q^2 p^2 (x^2-1)(1-y^2), \qquad (4.33)$$

we obtain the conjugate solution after some straightforward algebraic manipulations:

$$\hat{X} = \frac{\mu^2}{p^2}(1-y^2)\frac{[(1+px)^2+q^2]^2 - p^2 q^2 (x^2-1)(1-y^2)}{(1+px)^2+(qy)^2}, \qquad (4.34)$$

$$\hat{A} = -2\frac{qp}{\mu}\frac{(1+px)}{[(1+px)^2+q^2]^2 - p^2 q^2 (x^2-1)(1-y^2)}, \qquad (4.35)$$

$$e^{2\hat{h}} = \hat{X}\,\frac{(1+px)^2+(qy)^2}{p^2\,(x^2-y^2)}. \qquad (4.36)$$

The rotation potential $\hat{Y}$ ($d\hat{Y} = 2\hat{\omega}$) is computed by integrating $\hat{Y}_x = -[\mu(x^2-1)]^{-1}\hat{X}^2 \hat{A}_y$ and $\hat{Y}_y = [\mu(1-y^2)]^{-1}\hat{X}^2 \hat{A}_x$:

$$\hat{Y} = 2\frac{q\mu^2}{p^2}\,y\left[(3-y^2) + \frac{q^2(1-y^2)^2}{(1+px)^2+(qy)^2}\right]. \qquad (4.37)$$

Finally, the metric functions $\hat{W}$ and $\hat{V}$, which are given by $\hat{W} = \hat{A}\hat{X} = -AX$ and $\hat{V} = -X$, become

$$\hat{W} = -2\frac{\mu q}{p}\frac{(1-y^2)(1+px)}{(1+px)^2+(qy)^2} \qquad (4.38)$$

and

$$\hat{V} = -\frac{1-(px)^2-(qy)^2}{(1+px)^2+(qy)^2}. \qquad (4.39)$$

In terms of the conjugate quantities $\hat{V}$, $\hat{W}$, $\hat{X}$ and $\hat{h}$, the Kerr metric in prolate spheroidal coordinates reads

$$
\begin{aligned}
^{(4)}\boldsymbol{g} &= -\hat{V}\,dt^2 + 2\hat{W}\,dt d\varphi + \hat{X}\,d\varphi^2 \\
&\quad + \frac{e^{2\hat{h}}}{\hat{X}}\mu^2\,(x^2-y^2)\left[\frac{dx^2}{x^2-1}+\frac{dy^2}{1-y^2}\right]. \quad (4.40)
\end{aligned}
$$

For many purposes it is convenient to write the Kerr metric in terms of Boyer–Lindquist coordinates (t,r,ϑ,φ) (Boyer and Lindquist 1967), where r and ϑ are defined in terms of x and y by

$$r = m\,(1+px), \qquad \cos\vartheta = y. \qquad (4.41)$$

The parameters μ, p and q are replaced by the two new parameters m and a,

$$m = \mu\,p^{-1}, \qquad a = \mu\,q\,p^{-1}. \qquad (4.42)$$

(Note that $p^2+q^2=1$ implies $m^2-a^2=\mu^2$.) Introducing Δ and Ξ, defined by

$$\Delta = r^2 - 2mr + a^2, \qquad \Xi = r^2 + a^2\cos^2\vartheta, \qquad (4.43)$$

and using the relations

$$x^2-1 = \frac{\Delta}{\mu^2}, \qquad \frac{dx^2}{x^2-1} = \frac{1}{\Delta}\,dr^2, \qquad (4.44)$$

$$1-y^2 = \sin^2\vartheta, \qquad \frac{dy^2}{1-y^2} = d\vartheta^2, \qquad (4.45)$$

and the identities

$$(1+px)^2+(qy)^2 = m^{-2}\Xi, \qquad (1+px)^2+q^2 = m^{-2}(r^2+a^2), \qquad (4.46)$$

$$(px)^2 + (qy)^2 - 1 = m^{-2}\,(\Delta - a^2\sin^2\vartheta), \qquad (4.47)$$

$$1 + px = \frac{r}{m} = \frac{(r^2+a^2)-\Delta}{2\,m^2}, \qquad (4.48)$$

finally yields the result

$$\hat{X} = \Xi^{-1} \sin^2\vartheta \, [(r^2 + a^2)^2 - \Delta a^2 \sin^2\vartheta] , \qquad (4.49)$$

$$\hat{W} = \Xi^{-1} a \sin^2\vartheta \, [\Delta - (r^2 + a^2)] , \qquad (4.50)$$

$$\hat{V} = \Xi^{-1} [\Delta - a^2 \sin^2\vartheta] , \qquad (4.51)$$

$$e^{2\hat{h}} = \hat{X} \, \Xi \, [\Delta + (m^2 - a^2) \sin^2\vartheta]^{-1} , \qquad (4.52)$$

$$\hat{Y} = 2am \cos\vartheta \, [(3 - \cos^2\vartheta) + \Xi^{-1} a^2 \sin^4\vartheta] . \qquad (4.53)$$

In terms of Boyer–Lindquist coordinates, the Kerr metric assumes the form

$$^{(4)}g = \frac{1}{\Xi} [-(\Delta - a^2 \sin^2\vartheta) \, dt^2 + 2a \sin^2\vartheta \, (\Delta - (r^2 + a^2)) \, dt d\varphi$$

$$+ \sin^2\vartheta \, ((r^2 + a^2)^2 - \Delta a^2 \sin^2\vartheta) \, d\varphi^2] + \Xi \, [\frac{1}{\Delta} \, dr^2 + d\vartheta^2] . \quad (4.54)$$

The Kerr metric represents a 2–parameter family of stationary and axisymmetric black hole solutions of Einstein's vacuum equations. For $a = 0$ it reduces to the Schwarzschild metric with mass parameter m.

Corollary 4.4 *The quantities m and a parametrizing the Kerr metric (4.54) are identical to the total mass M and the ratio J/M, respectively, where J is the total angular momentum of the asymptotically flat, stationary and axisymmetric Kerr spacetime.*

Proof For large values of r one finds

$$(*(dr \wedge dt))_{\vartheta\varphi} = -\sqrt{|g|} \, g^{tt} g^{rr} = r^2 \sin\vartheta + \mathcal{O}(r) , \qquad (4.55)$$

$$(*(dr \wedge d\varphi))_{\vartheta\varphi} = -\sqrt{|g|} \, g^{t\varphi} g^{rr} = 2am \frac{\sin\vartheta}{r} + \mathcal{O}(\frac{1}{r^2}) . \qquad (4.56)$$

Using this in the expressions $k = -\hat{V} dt + \hat{W} d\varphi$ and $m = \hat{W} dt + \hat{X} d\varphi$ for the two Killing 1–forms, we obtain ($f' \equiv \partial f / \partial r$)

$$(*dk)_{\vartheta\varphi} = -\hat{V}' \, (*(dr \wedge dt))_{\vartheta\varphi} + \hat{W}' \, (*(dr \wedge d\varphi))_{\vartheta\varphi}$$

$$= -\frac{2m}{r^2} r^2 \sin\vartheta + \mathcal{O}(r^{-1}) ,$$

$$(*dm)_{\vartheta\varphi} = \hat{W}' \, (*(dr \wedge dt))_{\vartheta\varphi} + \hat{X}' \, (*(dr \wedge d\varphi))_{\vartheta\varphi}$$

$$= 2am \frac{\sin^2\vartheta}{r^2} r^2 \sin\vartheta + 4am \, r \sin^2\vartheta \frac{\sin\vartheta}{r} + \mathcal{O}(r^{-1}) .$$

The Komar expressions (2.34) and (2.73) for the total mass M and the total angular momentum J now yield

$$M = -\frac{1}{8\pi} \int_{S_\infty^2} *dk = \frac{2m}{8\pi} \int_{S_\infty^2} \sin\vartheta \, d\vartheta d\varphi = m \,, \qquad (4.57)$$

$$J = \frac{1}{16\pi} \int_{S_\infty^2} *dm = \frac{6am}{16\pi} \int_{S_\infty^2} \sin^3\vartheta \, d\vartheta d\varphi = a\,m. \qquad (4.58)$$

□

The asymptotic flatness of the Kerr metric is established from its expansion as $r \to \infty$:

$$^{(4)}g = -[1 - \frac{2m}{r} + \mathcal{O}(r^{-2})] \, dt^2 - [\frac{4am}{r} \sin^2\vartheta + \mathcal{O}(r^{-2})] \, dt \, d\varphi$$
$$+ [1 + \frac{2m}{r} + \mathcal{O}(r^{-2})] \, dr^2 + [1 + \mathcal{O}(r^{-1})] \, r^2 \, (d\vartheta^2 + \sin^2\vartheta d\varphi^2) \,.$$

This coincides with the expansion (2.33) of an asymptotically flat, stationary and axisymmetric metric in spherical coordinates and again confirms that $M = m$ and $J = ma$. For a rigorous proof of asymptotic flatness one has to verify the appropriate conditions for the conformally transformed metric $(1/r) \cdot {}^{(4)}g$ (see, e.g., Penrose 1963 or Ashtekar 1980). It is worth noting that the interpretation of r as a radial coordinate is not unproblematic since, in order to construct a complete spacetime, one has to allow for negative values of r as well. (In fact, the extended Kerr spacetime turns out to be also asymptotically flat as $r \to -\infty$.) This indicates that the Kerr spacetime exhibits some interesting, nontrivial features, for a detailed discussion of which we refer the reader to the literature (see, e.g., Hawking and Ellis 1973, Chandrasekhar 1983 or Wald 1984). Here we recall only the following facts:

(i) The singularity at $\Xi = 0$ ($r = 0$, $\vartheta = \pi/2$) is the only real curvature singularity of the Kerr metric (Boyer and Lindquist 1967, Carter 1968). Its nature is different from the corresponding singularity at the origin of the Schwarzschild or Reissner–Nordström spacetime, since it has the topology of a ring with coordinate radius a, as can be seen in Kerr–Schild coordinates (Kerr and Schild 1965). By analytic continuation of r to negative values, the Kerr spacetime can be extended to cover the region inside the ring as well (see Hawking and Ellis 1973).

(ii) The singularity at $\Xi = 0$ is "shielded" by a horizon if $M \geq |a|$, that is, if the equation $\Delta = 0$ has positive, real solutions. The latter occur for $r = r_\pm = M \pm (M^2 - a^2)^{1/2}$ and are coordinate singularities: $r = r_+$ being an event horizon and $r = r_-$ a Cauchy horizon.

(iii) In contrast to the Schwarzschild metric, where the horizon coincides with the surface $(k|k) = 0$, the latter is now located at $\Delta = a^2 \sin^2\vartheta$, that is, at $r = r_0(\vartheta) = M + (M^2 - a^2 \sin^2\vartheta)^{1/2}$. The only common points of the event horizon, $r = r_+$, and the ergoregion, $r = r_0(\vartheta)$, are the north pole, $\vartheta = 0$, and the south pole, $\vartheta = \pi/2$. Since the asymptotically timelike Killing field becomes spacelike for $r_+ < r < r_0(\vartheta)$, there exist no static observers in the ergoregion.

We conclude this chapter by recalling that a solution of the circular vacuum equations is determined by the Ernst potential. The uniqueness theorem for the Kerr metric will therefore be based on the uniqueness of regular, asymptotically flat exterior solutions to the Ernst equation, subject to the following boundary and regularity conditions (see Mazur 1984a, Carter 1985, Weinstein 1990):

(i) In prolate spheroidal coordinates, the asymptotic expansion (2.33) of a stationary and axisymmetric, asymptotically flat metric implies

$$\hat{X} = (1 - y^2)\,[\mu^2\,x^2 + \mathcal{O}(x)], \qquad (4.59)$$

$$\hat{Y} = 2\,J\,y\,(3 - y^2) + \mathcal{O}(x^{-1}). \qquad (4.60)$$

Here we have used eq. (4.17) to obtain the asymptotic behavior of $\hat{Y}$ from the expansion of $\hat{A}$. (Also note that $(\mu/p)(\mu q/p) = ma = J$.) These formulae are in accordance with the expansions of the Kerr potentials (4.34) and (4.37) for $x \to \infty$.

(ii) In the vicinity of the rotation axis, $y \to \pm 1$, we require

$$\hat{X} = \mathcal{O}(1 - y^2), \quad (1 - y^2)\,\frac{\partial_y\hat{X}}{\hat{X}} = \mp 2 + \mathcal{O}(1 - y^2), \quad (4.61)$$

$$\hat{Y} = \pm 4\,J + \mathcal{O}(1 - y^2), \quad \partial_x\hat{Y} = \mathcal{O}((1 - y^2)^2), \quad (4.62)$$

which is also fulfilled by the Kerr solution (4.34), (4.37). In particular, $\partial_y\hat{X}/\hat{X}$ remains finite, despite the fact that $\hat{X}$ vanishes on the axis.

(iii) Finally, the solution (4.34), (4.37) shows that the potentials $\hat{X}$ and $\hat{Y}$ behave regularly on the horizon, $x = 1$.

5

Electrovac spacetimes with Killing fields

In this chapter we consider self–gravitating electromagnetic fields which are invariant under the action of one or more Killing fields. As an application, we present the derivation of the Kerr–Newman metric in the last section.

In the first section we introduce the electric and magnetic 1–forms, $E = -i_K F$ and $B = i_K * F$, which can be defined in terms of the electromagnetic field tensor (2–form) F and a Killing field K. We also express the stress–energy tensor in terms of K, E and B. We then focus on the case where spacetime admits two Killing fields, k and m, say, and establish some algebraic identities between their associated electric and magnetic 1–forms.

The invariance conditions for electromagnetic fields are introduced in the second section. The homogeneous Maxwell equations imply the existence of a complex potential, Λ, which can be associated with E and B if the Killing field K acts as a symmetry transformation on F. In terms of Λ, the remaining Maxwell equations reduce to one complex equation for $\Delta\Lambda$, involving the twist, ω, and the norm, N, of the Killing field. In the Abelian case, to which we restrict our attention in this chapter, the presence of gauge freedom does not require a modified invariance concept for F. In contrast, the symmetry conditions must be reexamined if one deals with arbitrary gauge groups (see, e.g., Forgács and Manton 1980, Harnad et al. 1980, Jackiw and Manton 1980, Brodbeck and Straumann 1993, 1994, Heusler and Straumann 1993b).

Taking advantage of the general relations derived in section 2.2, we introduce the electrovac Ernst potential E in the third section. The real part of E is composed of the norm of the Killing field and $|\Lambda|^2$, whereas the imaginary part combines the twist and the 1–form $(\overline{\Lambda}d\Lambda - \Lambda\overline{d\Lambda})$ into a common potential. In terms of E and Λ, the $R(K)$–Einstein equations and Maxwell's equations reduce to two complex equations for ΔE and $\Delta\Lambda$. As in the vacuum case (see section 2.3), these relations can also be obtained from a variational principle.

In the presence of a second Killing field, the symmetry conditions imply that the components $F(k,m)$ and $(*F)(k,m)$ of the field tensor vanish. As we shall argue in the fourth section, this enables one to conclude that spacetime is Ricci–circular. The circularity theorem 2.20 then implies that the Killing fields are integrable.

In the fifth section we use the general properties derived in chapter 3 for circular spacetimes. By virtue of these, the Einstein–Maxwell equations reduce to a *decoupled* system of Ernst equations for E and Λ, and two additional equations for the remaining metric functions in terms of these potentials.

The last section is devoted to a detailed derivation of the Kerr–Newman metric (Newman *et al.* 1965). Introducing the potentials $\epsilon \equiv (1 + E)/(1 - E)$ and $\lambda \equiv 2\Lambda/(1 - E)$, we show that the Kerr–Newman solution is obtained from the linear ansatz $\lambda = \lambda_0$, $\epsilon = px + iqy$, where the constant λ_0 is related to the electric charge (and x, y are prolate spheroidal coordinates). As in the vacuum case, the remaining metric functions can be computed from the Ernst potentials by quadrature. The Kerr–Newman metric is finally obtained by considering the conjugate solution.

5.1 The stress–energy tensor

In this section we introduce the electric and magnetic 1–forms, E and B, which can be associated with the electromagnetic field tensor F and a Killing field K. In particular, we give the expression for the stress–energy tensor in terms of these quantities and derive some algebraic identities between the 1–forms associated with two different Killing fields.

Consider the electromagnetic action

$$S[A] = \int_M \mathcal{L}[A]\,\eta = \frac{1}{8\pi} \int_M F \wedge *F \qquad (5.1)$$

(with $\mathcal{L}[A] = (8\pi)^{-1}(F|F) = (16\pi)^{-1}F_{\mu\nu}F^{\mu\nu}$), where $F = dA$ denotes the field strength 2–form. The stress–energy tensor is obtained from the variation

$$\delta_g \, S[A] \ = \ -\frac{1}{2} \int_M T^{\mu\nu} \, \delta^{\,(4)} g_{\mu\nu} \, \eta \,, \qquad (5.2)$$

which yields

$$T_{\mu\nu} \ = \ 2 \frac{\partial \mathcal{L}}{\partial^{\,(4)} g^{\mu\nu}} \ - \ ^{(4)} g_{\mu\nu} \, \mathcal{L} \,, \qquad (5.3)$$

since the variation with respect to the metric involves no integrations by parts, that is, $\mathcal{L}[A]$ does not depend on connection coefficients. (δ_g denotes variations with respect to the metric.) The electromagnetic stress–energy tensor becomes

$$T_{\mu\nu} \ = \ \frac{1}{4\pi} \, [\, F_{\mu\sigma} \, F_{\nu}^{\ \sigma} \ - \ \frac{1}{2} \, ^{(4)} g_{\mu\nu} \, (F|F) \,] \,, \qquad (5.4)$$

(with $\operatorname{tr} \boldsymbol{T} =^{(4)} g^{\mu\nu} T_{\mu\nu} = (4\pi)^{-1}[F_{\mu}^{\ \sigma} F^{\mu}_{\ \sigma} - 2(F|F)] = 0$).

For an arbitrary Killing field K, we define the 1–forms $E \in \Lambda_1(M)$ and $B \in \Lambda_1(M)$ by

$$E \ = \ -i_K \, F \ = \ *(K \wedge *F), \qquad (5.5)$$

$$B \ = \ i_K * F \ = \ *(K \wedge F). \qquad (5.6)$$

We shall refer to E and B as the electric and the magnetic components of F with respect to K. The electromagnetic 2–form, F, is recovered from E, B and K by the relation

$$F \ = \ -\frac{1}{N} \, [\, K \wedge E \ + \ *(K \wedge B) \,] \,, \qquad (5.7)$$

where, as usual, $N = (K|K)$ denotes the norm of the Killing field. (To verify the above equation one applies i_K and $i_K *$ to it and uses the consequences $i_K E = 0$ and $i_K B = 0$ of the definitions (5.5) and (5.6): $i_K * F = -N^{-1} i_K *^2 (K \wedge B) = N^{-1} i_K (K \wedge B) = B$.) Using the general identity (1.25), we also obtain $(K|K)(F|F) = (i_K F | i_K F) - (i_K * F | i_K * F)$, and thus

$$(F|F) \ = \ \frac{1}{N}[(E|E) \ - \ (B|B)]. \qquad (5.8)$$

The following expression for the stress–energy tensor in terms of E, B and K turns out to be very useful:

$$4\pi \, \boldsymbol{T} \ = \ \frac{1}{N^2} \, [(E|E) + (B|B)] \, \{ K \otimes K - \frac{N}{2} \, ^{(4)} \boldsymbol{g} \}$$

$$+ \frac{1}{N} \{ E \otimes E + B \otimes B \} + \frac{1}{N^2} \{ K \otimes S + S \otimes K \}, \qquad (5.9)$$

where S is proportional to the Poynting vector (1–form) as seen by an observer moving along the Killing trajectories (Rácz 1993),

$$S = *(K \wedge E \wedge B). \tag{5.10}$$

The derivation of the above formula is left to the reader as an exercise. Hint: Consider two arbitrary vector fields, X and Y, say, and write $4\pi T(X,Y) = (i_X F | i_Y F) - (1/2)(X|Y)(F|F)$. Use eq. (5.8) in the second term and compute the first term by taking advantage of eq. (5.7) and the identities given at the end of chapter 1. This results in the expressions: $-N i_X F = (K|X)E - (E|X)K + *(X \wedge K \wedge B)$; $N^2(i_X F | i_Y F) = (K|X)(K|Y)[(E|E) + (B|B)] - N(X|Y)(B|B) + N[(X|B)(Y|B) + (X|E)(Y|E)] + (K|X)(K \wedge Y | * [E \wedge B]) + (K|Y)(K \wedge X | * [E \wedge B])$.

As an application of eq. (5.9) one easily derives the following expression for the electromagnetic 1–form $T(K)$ (with components $T(K)_\mu = T_{\mu\nu}K^\nu$):

$$8\pi\, T(K) = [(E|E) + (B|B)]\frac{K}{N} + 2 * (\frac{K}{N} \wedge E \wedge B). \tag{5.11}$$

By considering the interior and the exterior product with K, this also yields the identities

$$T(K,K) = \frac{1}{8\pi}[(E|E) + (B|B)], \tag{5.12}$$

$$*(K \wedge T(K)) = -\frac{1}{4\pi}(E \wedge B), \tag{5.13}$$

where $T(K,K) \equiv i_K T(K) = T_{\mu\nu}K^\mu K^\nu$. The above formulae imply the identity

$$\boldsymbol{T} + \frac{1}{N}T(K,K)\,^{(4)}\boldsymbol{g} - \frac{1}{N}\{K \otimes T(K) + T(K) \otimes K\}$$

$$= \frac{1}{4\pi N}\{E \otimes E + B \otimes B\}, \tag{5.14}$$

which bears a striking resemblance to the formula (2.20) for the projection of the Ricci tensor in the presence of a Killing field. Using the two formulae (i.e., eqs. (5.14) and (2.20)) in Einstein's equations with electromagnetic fields, $\boldsymbol{R} = 8\pi\,\boldsymbol{T}$, we immediately find

$$\boldsymbol{R}^{(P)} = \frac{1}{2N^2}\{dN \otimes dN + 4\,\omega \otimes \omega\} + \frac{2}{N}\{E \otimes E + B \otimes B\}, \tag{5.15}$$

where $\boldsymbol{R}^{(P)}$ is the Ricci tensor with respect to the projection metric $\boldsymbol{P} = |N|(^{(4)}\boldsymbol{g} - N^{-1}K \otimes K)$ (see eq. (2.19)). The similarity of eqs. (5.14) and (2.20) is also responsible for the fact that the vacuum variational principle (section 2.3) generalizes to electrovac spacetimes (see below).

Later we shall be interested in stationary and axisymmetric spacetimes. We therefore also derive some identities involving *two* Killing fields. The following notations are used throughout: The electric and magnetic 1–forms obtained from the asymptotically timelike Killing field k are denoted by E_k and B_k, respectively, whereas E_m and B_m are used for the corresponding quantities associated with the axial Killing field m:

$$E_k = -i_k F, \qquad B_k = i_k * F, \qquad (5.16)$$

$$E_m = -i_m F, \qquad B_m = i_m * F. \qquad (5.17)$$

The following algebraic relations between E_k, B_k, E_m and B_m turn out to be useful:

$$W * F = m \wedge B_k - *(k \wedge E_m) = k \wedge B_m - *(m \wedge E_k), \quad (5.18)$$

$$X B_k - W B_m = (*F)(k, m) m - *(\Omega \wedge E_m), \quad (5.19)$$

$$X E_k - W E_m = -F(k, m) m + *(\Omega \wedge B_m), \quad (5.20)$$

where $\Omega = k \wedge m$ denotes the Killing 2–form defined in eq. (2.79), and $X = (m|m)$, $W = (m|k)$ and $V = -(k|k)$ were introduced in equation (3.10). The relation (5.18) is a consequence of the definitions (5.16) and (5.17) since, for instance,

$$*(k \wedge E_m) = -i_k * E_m = -i_k(m \wedge *F) = -W * F + m \wedge B_k.$$

In order to derive the expression (5.19) for $X B_k$, one applies i_m to the first part of eq. (5.18). This yields

$$\begin{aligned} W B_m &= i_m W * F = i_m(m \wedge B_k) - i_m *(k \wedge E_m) \\ &= X B_k - (m|B_k) m - *(m \wedge k \wedge E_m), \end{aligned}$$

where $(m|B_k) = i_m i_k * F = (*F)(k, m)$. In a similar way one obtains eq. (5.20) for $X E_k$, applying i_m to the dual of the second part of the identity (5.18). Using these identities and eq. (5.11) one also finds the expression

$$\begin{aligned} 8\pi T(k, m) &= \frac{W}{X}[(E_m|E_m) + (B_m|B_m)] - \frac{2}{X}(\Omega| * [E_m \wedge B_m]) \\ &= (E_k|E_m) + (B_k|B_m). \end{aligned} \qquad (5.21)$$

for the mixed components of the stress–energy tensor. Having established the electrovac circularity theorem, eqs. (5.19) and (5.20) will also provide us with the compact expression (5.58), which turns out to be crucial in order to discuss the relationship between conjugate solutions.

The following proposition is important for the derivation of the Kerr–Newman solution. Together with the circularity theorem for electrovac spacetimes (see the fourth section), it implies that $\mathrm{tr}_\sigma R = 0$ (see eq. (3.33)). As in the vacuum case, this will be used to establish that $\sqrt{-\sigma}$ is a harmonic function.

Proposition 5.1 *Let k and m be two Killing fields and let $T_{\mu\nu}$ be the electromagnetic stress–energy tensor. Then*

$$8\pi[\,(k|k)\,T(m,m)\,+\,(m|m)\,T(k,k)\,-\,2\,(k|m)\,T(k,m)\,]$$
$$= 2\,[F(k,m)]^2\,+\,2\,[(*F)(k,m)]^2\,. \tag{5.22}$$

Proof Taking advantage of eq. (5.12) and the identity (1.26), we obtain for the terms on the l.h.s. of eq. (5.22),

$$(k|k)\,[(E_m|E_m) + (B_m|B_m)]\,+\,(m|m)\,[(E_k|E_k) + (B_k|B_k)]$$
$$-2\,(k|m)\,[(E_k|E_m) + (B_k|B_m)]\,=\,2\,(i_k i_m F)^2 + 2\,(i_k i_m * F)^2$$
$$+\,|\,k \wedge E_m - m \wedge E_k\,|^2\,+\,|\,k \wedge B_m - m \wedge B_k\,|^2\,,$$

where $|\alpha|^2 \equiv (\alpha|\alpha)$. Here we have again used the consequences $(k|E_k) = (k|B_k) = 0$ and $(m|E_m) = (m|B_m) = 0$ of the definitions (5.16) and (5.17). The entire last line vanishes since $k \wedge E_m - m \wedge E_k$ is the dual of $k \wedge B_m - m \wedge B_k$ (see eq. (5.18)) and since $(\alpha|\alpha) + (*\alpha| * \alpha) = 0$ for any p–form α.   □

5.2 Maxwell's equations with symmetries

So far, we have established a set of *algebraic* identities between the quantities which can be formed from F and two Killing fields, k and m, say. In order to proceed, we now consider the invariance conditions for the electromagnetic field F with respect to the action of a Killing field. We will then discuss the implications of these conditions in combination with Maxwell's and Einstein's equations. This will lead us to the staticity and circularity issues for electrovac spacetimes, which we shall address in the fourth section of this chapter.

Let K denote an arbitrary Killing field. The electromagnetic field is invariant under the action generated by K if the Lie derivative of F with respect to K vanishes. Since the Lie derivative with respect to a Killing vector field commutes with the Hodge dual (see eq. (2.3)), this implies that the dual 2–form $*F$ is invariant as well,

$$L_K F = 0, \quad L_K * F = 0. \tag{5.23}$$

Definition 5.2 *An electromagnetic field is called stationary if it is invariant under the 1–parameter group of transformations generated by the (asymptotically) timelike Killing field k, $L_k F = 0$. If F is invariant under the 1–parameter group generated by an axial Killing field m, $L_m F = 0$, then the electromagnetic field is called axisymmetric.*

It is worthwhile pointing out that the above definition needs to be modified for Yang–Mills fields. As a matter of fact, even in the Abelian case, the invariance condition $L_K A = 0$ for the gauge *potential* is *not* gauge invariant. It must be replaced by $L_K A = d\mathcal{V}$, where $\mathcal{V}$ is the generator of an infinitesimal gauge transformation (see, e.g., Forgács and Manton 1980, Heusler and Straumann 1993b). However, in the Abelian case, the above invariance condition for A yields $L_K F = 0$, since $L_K dA = dL_K A = d^2 \mathcal{V} = 0$. In contrast to this, the corresponding non–Abelian condition, $L_K A = D\mathcal{V}$, implies the modified symmetry condition $L_K F = [F, \mathcal{V}]$ since

$$
\begin{aligned}
L_K F &= L_K DA = L_K (dA + A \wedge A) \\
&= dL_K A + [A, L_K A] = DL_K A = D^2 \mathcal{V} = [F, \mathcal{V}].
\end{aligned}
$$

Although it is possible to choose a gauge where $\mathcal{V} = 0$, it is appropriate to describe the invariance properties for Yang–Mills fields in a gauge–invariant manner. In particular, adopting a gauge where $\mathcal{V}$ vanishes restricts the freedom to impose boundary conditions for the electric potential. In fact, this problem already occurs in the Abelian case, if Higgs fields are also taken into account (see Gibbons 1991). For the pure Einstein–Maxwell system, to which we restrict our attention in this chapter, it is, however, sufficient to adopt the notion of symmetry given in the above definition.

The following proposition gives the Maxwell equations in terms of the electric and magnetic 1–forms:

Proposition 5.3 *Consider an electromagnetic field $F \in \Lambda_2(M)$ and a current $j \in \Lambda_1(M)$, subject to the Maxwell equations*

$$dF = 0, \quad d * F = 4\pi * j. \qquad (5.24)$$

Let F be invariant under the action of the Killing field K. Then the 1–forms $E = -i_K F$ and $B = i_K * F$ fulfil the following equations:

$$\begin{aligned}
dE &= 0, \\
dB &= 4\pi * (K \wedge j), \qquad (5.25) \\
d^\dagger \left(\frac{E}{N} \right) &= 2 \frac{(\omega|B)}{N^2} + 4\pi \frac{(K|j)}{N}, \\
d^\dagger \left(\frac{B}{N} \right) &= -2 \frac{(\omega|E)}{N^2}, \qquad (5.26)
\end{aligned}$$

where N and ω denote the norm and the twist of K, respectively, $N \equiv (K|K)$ and $\omega \equiv (1/2) * (K \wedge dK)$.

Proof Using $L_K = i_K d + d i_K$, the invariance condition (5.23) and Maxwell's equations (5.24), we have

$$\begin{aligned}
dE &= -d i_K F = -L_K F + i_K dF = 0, \\
dB &= d i_K * F = L_K * F - i_K d * F = 4\pi * (K \wedge j).
\end{aligned}$$

As an immediate consequence of these equations and $(K|E) = (K|B) = 0$, we also obtain the invariance of E and B under the action of K,

$$L_K E = 0, \quad L_K B = 0, \qquad (5.27)$$

since, e.g., $L_K B = i_K dB + d(K|B) = 4\pi * (k \wedge k \wedge j) = 0$. The co–derivative equations (5.26) are obtained as follows: We first use $d^\dagger = *d*$ and $E = -i_K F = *(K \wedge *F)$ to write

$$d^\dagger \left(\frac{E}{N} \right) = *d \left[\frac{K}{N} \wedge *F \right] = * \left[d \left(\frac{K}{N} \right) \wedge *F \right] - * \left[\frac{K}{N} \wedge d * F \right].$$

The second term has already the desired form since

$$- * \left[\frac{K}{N} \wedge d * F \right] = -4\pi * \left[\frac{K}{N} \wedge *j \right] = 4\pi \frac{(K|j)}{N},$$

where we have used $*\eta = -1$. In the first term we use the general identity (2.12) for $d(K/N)$ and the expression (5.7) for F in terms of E and B. Now taking advantage of $(\alpha|\beta) = -(*\alpha| * \beta)$ and the fact that the electric part of $*F$ does not contribute, we find

$$*[d\left(\frac{K}{N}\right)\wedge *F] = (*d\left(\frac{K}{N}\right)|*F) = 2(\frac{K\wedge\omega}{N^2}|\frac{K\wedge B}{N}) = \frac{2}{N^2}(\omega|B),$$

which completes the proof of the first relation in eq. (5.26). In a similar manner one establishes the formula for the co–derivative of (B/N). $\qquad\square$

The following corollary shows that the equations (5.25) and (5.26) for E and B are equivalent to the Poisson equation for one complex potential:

Corollary 5.4 *Let F be invariant under the action of a Killing field K and let $K \wedge j = 0$. Then there exists (locally) a potential Λ, such that*

$$d\Lambda = -E + iB = i_K(F + i * F) \qquad (5.28)$$

and

$$\Delta\Lambda = -\frac{(d\Lambda| - dN + 2i\,\omega)}{N} + 4\pi\,(K|j). \qquad (5.29)$$

In terms of Λ, $T(K,K)$ and $K \wedge T(K)$ become

$$T(K,K) = \frac{1}{8\pi}(d\Lambda|\overline{d\Lambda}), \qquad (5.30)$$

$$*(K\wedge T(K)) = \frac{i}{16\pi}d(\Lambda\overline{d\Lambda} - \overline{\Lambda}d\Lambda). \qquad (5.31)$$

Proof Clearly $dE = dB = 0$ implies the local existence of a complex potential Λ with $d\Lambda$ according to eq. (5.28). Using $\Delta\Lambda = -d^\dagger d\Lambda$, the Maxwell equations (5.26) imply eq. (5.29), since

$$\Delta\Lambda = 2\frac{(\omega|B + iE)}{N} - \frac{(dN|E - iB)}{N} + 4\pi\,(K|j).$$

The expressions (5.30) for $T(K,K)$ and (5.31) for $*(K\wedge T(K))$ can be read off from the corresponding equations (5.12) and (5.13) and the definition (5.28). $\qquad\square$

5.3 The electrovac variational principle

In section 2.2 we derived the basic relations between the twist 1–form, ω, the norm, N, and the Ricci 1–form, $R(K)$, assigned to an arbitrary Killing field K (see eqs. (2.16)-(2.18)). In the vacuum case, these identities imply the existence of the complex Ernst potential E, which satisfies the differential equation (2.25). (In the presence of a second Killing field this equation decouples and becomes the Ernst equation (4.6) on a 2–dimensional manifold.) The purpose of this section is to show that the system (2.16)-(2.18), together with Maxwell's equation (5.29) and Einstein's equations for $R(K, K)$ and $K \wedge R(K)$, implies the existence of a modified Ernst potential E. In addition, we shall see that the entire set of Einstein–Maxwell equations with a Killing field can be obtained from a variational principle for E, Λ and the projection metric $\boldsymbol{P}$ defined in eq. (2.19).

We start by using Einstein's equations

$$R(K, K) = 8\pi \left[T(K, K) - \frac{1}{2} N \operatorname{tr} \boldsymbol{T} \right],$$

$$* (K \wedge R(K)) = 8\pi * (K \wedge T(K)), \tag{5.32}$$

(with $\operatorname{tr} \boldsymbol{T} = 0$) and eqs. (5.30), (5.31) to write the identities (2.16)-(2.18) and the Maxwell equation (5.29) in the form

$$d \left(2i\,\omega + \Lambda \overline{d\Lambda} - \overline{\Lambda} d\Lambda \right) = 0, \tag{5.33}$$

$$d^{\dagger} (2i\,\omega) = - \frac{(4i\,\omega | dN)}{N}, \tag{5.34}$$

$$\Delta N = \frac{(dN|dN) - 4\,(\omega|\omega)}{N} - 2\,(d\Lambda|\overline{d\Lambda}), \tag{5.35}$$

$$\Delta \Lambda = - \frac{(d\Lambda| - dN + 2i\,\omega)}{N} + 4\pi\,(K|j). \tag{5.36}$$

Despite the fact that $d\omega$ does not vanish, eq. (5.33) still implies the existence of a potential Y with

$$dY = 2\omega - i\,(\Lambda \overline{d\Lambda} - \overline{\Lambda} d\Lambda), \tag{5.37}$$

which reduces to the vacuum twist potential when $\Lambda = 0$. Now defining the Ernst potential E by

$$\mathrm{E} = - (N + \Lambda \overline{\Lambda}) + iY, \tag{5.38}$$

we obtain from eq. (5.37):

$$d\mathrm{E} = - dN + 2i\,\omega - 2\overline{\Lambda} d\Lambda. \tag{5.39}$$

Together with eqs. (5.34)-(5.36), this yields the following expression for the Laplacian of E:

$$\Delta E = -\Delta N - 2\,(d\Lambda|\overline{d\Lambda}) - 2\,\overline{\Lambda}\Delta\Lambda - d^\dagger(2i\,\omega)$$

$$= \frac{4(\omega|\omega) - (dN|dN)}{N} + \frac{(2\overline{\Lambda}d\Lambda| - dN + 2i\omega)}{N} + \frac{(4i\omega|dN)}{N}$$

$$= -\frac{(dE|dE + 2\overline{\Lambda}d\Lambda)}{N}.$$

Here we have restricted ourselves to the source–free case, $j = 0$. To summarize, we have the following corollary:

Corollary 5.5 *Let F be invariant with respect to the action of a Killing field K with norm N and twist ω. Then the electrovac Einstein–Maxwell equations imply the (local) existence of the two potentials Λ and E, such that*

$$d\Lambda = i_K\,(F + i * F), \quad dE = -\,dN + 2i\,\omega - 2\,\overline{\Lambda}d\Lambda. \quad (5.40)$$

The potentials are subject to the equations

$$\Delta\Lambda = -\frac{(d\Lambda|dE + 2\overline{\Lambda}d\Lambda)}{N(E,\Lambda)}, \quad \Delta E = -\frac{(dE|dE + 2\overline{\Lambda}d\Lambda)}{N(E,\Lambda)}, \quad (5.41)$$

which are also the Euler–Lagrange equations for the action

$$S[E, \overline{E}, \Lambda, \overline{\Lambda}] = \int_M * \left[\frac{(dE + 2\overline{\Lambda}d\Lambda|\overline{dE + 2\overline{\Lambda}d\Lambda})}{N^2(E,\Lambda)} + 4\frac{(d\Lambda|\overline{d\Lambda})}{N(E,\Lambda)} \right], \quad (5.42)$$

where

$$N = N(E, \Lambda) = -Re(E) - \Lambda\overline{\Lambda}. \quad (5.43)$$

Proof It remains to prove the last assertion: Making use of

$$dN = -\frac{1}{2}\,(dE + 2\overline{\Lambda}d\Lambda) - \frac{1}{2}\,\overline{(dE + 2\overline{\Lambda}d\Lambda)}$$

and $\delta_{\overline{E}}N = -\frac{1}{2}\delta\overline{E}$, we obtain

$$\delta_{\overline{E}}\,S[E, \overline{E}, \Lambda, \overline{\Lambda}] = \int_M d\delta\overline{E} \wedge \frac{*(dE + 2\overline{\Lambda}d\Lambda)}{N^2}$$

$$- \int_M \delta_{\overline{E}}N \left[2\frac{*(dE + 2\overline{\Lambda}d\Lambda|\overline{dE + 2\overline{\Lambda}d\Lambda})}{N^3} + 4\frac{*(d\Lambda|\overline{d\Lambda})}{N^2} \right]$$

$$\doteq \int_M \frac{\delta\overline{E}}{N^2}\left[-d*(dE+2\overline{\Lambda}d\Lambda)-\frac{*(dE+2\overline{\Lambda}d\Lambda|dE+2\overline{\Lambda}d\Lambda)}{N}\right]$$

$$+\int_M \frac{\delta\overline{E}}{N^2}2*(d\Lambda|\overline{d\Lambda})=-\int_M \frac{\delta\overline{E}}{N^2}*\left[\Delta E+\frac{(dE|dE+2\overline{\Lambda}d\Lambda)}{N}\right]$$

$$-\int_M 2\frac{\delta\overline{E}}{N^2}\,\overline{\Lambda}*\left[\Delta\Lambda+\frac{(d\Lambda|dE+2\overline{\Lambda}d\Lambda)}{N}\right],$$

where '$\doteq$' stands for equal up to boundary terms which arise from integrations by parts. A similar equation, involving a different combination of the two last integrals, is obtained from variations with respect to $\overline{\Lambda}$. Requiring that both $\delta_{\overline{E}}S$ and $\delta_{\overline{\Lambda}}S$ vanish yields the Euler–Lagrange equations (5.41). $\qquad\Box$

It is important to note that the Lagrangian assigned to the action (5.42) is manifestly positive definite if N is positive. It is mainly for this reason that we shall formulate the stationary and axisymmetric Ernst system on the basis of the spacelike Killing field m. If, instead, we were to choose the asymptotically timelike field k as the fundamental Killing field, the two integrands in the action (5.42) would not assume the same sign in the whole domain of outer communications.

We conclude this section by noting that - as in the vacuum case - the remaining Einstein equations are also obtained from a variational principle: First, the electrovac projection formula (5.15) can be written in terms of the complex potentials as

$$\boldsymbol{R}^{(P)}=\frac{1}{2}\frac{(dE+2\overline{\Lambda}d\Lambda)\otimes\overline{(dE+2\overline{\Lambda}d\Lambda)}}{N^2}+2\frac{d\Lambda\otimes\overline{d\Lambda}}{N}.\quad(5.44)$$

Since the variation of $\boldsymbol{R}^{(P)}$ does not depend on the Ernst potentials, and since $\delta(\sqrt{|P|}P^{\mu\nu}R_{\mu\nu}^{(P)})\doteq R_{\mu\nu}^{(P)}\delta(\sqrt{|P|}P^{\mu\nu})$, the effective Lagrangian becomes

$$L[E,\overline{E},\Lambda,\overline{\Lambda},P]=\sqrt{|P|}\,P^{\mu\nu}\times$$

$$\left[R_{\mu\nu}^{(P)}-\left(\frac{(E_{,\mu}+2\overline{\Lambda}\Lambda_{,\mu})(\overline{E}_{,\nu}+2\Lambda\overline{\Lambda}_{,\nu})}{2\,N^2(E,\Lambda)}+2\frac{\Lambda_{,\mu}\,\overline{\Lambda}_{,\nu}}{N(E,\Lambda)}\right)\right],\quad(5.45)$$

which generalizes the corresponding vacuum expression given in eq. (2.28).

Table 5.1. *Symmetry Transformations*

$(E, \Lambda) \mapsto \tilde{E} =$	$(E, \Lambda) \mapsto \tilde{\Lambda} =$
$\|\alpha\|^2 E$	$\alpha \Lambda$
$E + iu$	Λ
$E (1 + ivE)^{-1}$	$\Lambda (1 + ivE)^{-1}$
$E - 2\bar{\beta}\Lambda - \|\beta\|^2$	$\Lambda + \beta$
$E (1 - 2\bar{\gamma}\Lambda - \|\gamma\|^2 E)^{-1}$	$(\Lambda + \gamma E) (1 - 2\bar{\gamma}\Lambda - \|\gamma\|^2 E)^{-1}$

The above discussion shows that the Einstein–Maxwell equations in the presence of *one* Killing field are the Euler–Lagrange equations for the two potentials E, Λ and the 3–metric *P*. This enables one to generate new solutions from given ones, by finding the invariance transformations of the Lagrangian (5.45). Kinnersley (1973) has shown that this can be achieved by parametrizing the Ernst potentials E and Λ by the complex 3–vector v,

$$(v_0, v_1, v_2) = \frac{1}{2\sqrt{|N|}} (E - 1, E + 1, 2\Lambda). \qquad (5.46)$$

In terms of the Kinnersley vector v, the action (5.42) exhibits a manifest $SU(2,1)$ invariance (Kinnersley 1977, Kinnersley and Chitre 1977, 1978a, 1978b, Hoenselars *et al.* 1979; see also section 10.2). In the vacuum case, the symmetry group is $SU(1,1)$, as had been pointed out earlier by Geroch (1971). The associated invariance transformations for E and Λ are parametrized by eight real parameters, $\alpha, \beta, \gamma \in \mathbb{C}$; $u, v \in \mathbb{R}$, and are given in table 5.1. They provide a nonlinear representation of the Kinnersley group, and were found by Neugebauer and Kramer (1969) by analyzing the Killing vectors of the potential space (i.e., the 4–dimensional target manifold with coordinates Re(E), Im(E), Re(Λ), Im(Λ) and metric given by the integrand in eq. (5.42)). Here we shall not go into the details of the generation techniques; a comprehensive account and references to the literature can be found in Kramer *et al.* (1980).

5.4 The electrovac circularity theorem

In this section we consider electromagnetic fields which are invariant under the action of both the Killing field k generating the stationary isometry and the Killing field m generating the axial isometry. In section 2.7 we quoted a theorem stating that k and m commute with each other if spacetime is asymptotically flat. By virtue of theorem 2.20, the integrability conditions for k and m are fulfilled if spacetime is Ricci–circular. It therefore remains to establish the Ricci–circularity conditions. This is achieved by taking advantage of the field equations and the symmetry conditions for the electromagnetic 2–form F. We start by proving the following proposition:

Proposition 5.6 *Consider an asymptotically flat, stationary and axisymmetric spacetime with Killing fields k and m. Let F and j denote the stationary and axisymmetric electromagnetic field and current, respectively;*

$$L_k F = L_m F = 0 , \qquad k \wedge m \wedge j = 0 . \qquad (5.47)$$

Then, in every domain of spacetime which intersects the rotation axis, Maxwell's equations imply

$$F(k, m) = 0 , \qquad (*F)(k, m) = 0 . \qquad (5.48)$$

Proof Commuting Killing fields fulfil the identity

$$d \, i_m \, i_k = i_m \, i_k \, d + i_k \, L_m - i_m \, L_k , \qquad (5.49)$$

since $d i_m i_k = (L_m - i_m d) i_k = L_m i_k - i_m (L_k - i_k d)$. Using Maxwell's equations (5.24) and the commutation property of the Hodge dual with the Lie–derivative with respect to a Killing field (see eq. (2.3)), we obtain

$$d \, i_m \, i_k \, F = (i_k \, L_m - i_m \, L_k) \, F , \qquad (5.50)$$

$$d i_m i_k * F = (i_k * L_m - i_m * L_k) F + 4\pi * (k \wedge m \wedge j). \qquad (5.51)$$

Hence, by virtue of the symmetry conditions (5.47), both 1–forms $d[F(k, m)]$ and $d[(*F)(k, m)]$ vanish, implying that the functions $F(k, m)$ and $(*F)(k, m)$ assume constant values and vanish on the rotation axis (where $m = 0$). $\qquad \square$

Using the Einstein equations, $k \wedge R(k) = 8\pi k \wedge T(k)$ and $m \wedge R(m) = 8\pi m \wedge T(m)$, and the algebraic identity (5.13), it is now easy to verify that $F(k,m) = 0$ and $(*F)(k,m) = 0$ together imply Ricci–circularity (2.77):

$$i_m * (k \wedge R(k)) = -2i_m(E_k \wedge B_k)$$
$$= 2\left[F(k,m)\,B_k + (*F)(k,m)\,E_k\right], \qquad (5.52)$$
$$i_k * (m \wedge R(m)) = -2i_k(E_m \wedge B_m)$$
$$= -2\left[F(k,m)\,B_m + (*F)(k,m)\,E_m\right], \qquad (5.53)$$

where we have used $(m|E_k) = -i_m i_k F = -F(k,m) = i_k i_m F = -(k|E_m)$ and $(m|B_k) = (*F)(k,m) = -(k|B_m)$. The above proposition now demonstrates that an asymptotically flat, stationary and axisymmetric electrovac spacetime is Ricci–circular. As a corollary to theorem 2.20 one obtains the electromagnetic generalization (Carter 1969) of the vacuum Kundt and Trümper (1966) theorem:

Theorem 5.7 *Consider an asymptotically flat, stationary and axisymmetric spacetime containing a stationary and axisymmetric electromagnetic field (and current j, with $k \wedge m \wedge j = 0$). Then every domain of spacetime intersecting the rotation axis is circular,*

$$(m|\omega_k) = 0\,, \qquad (k|\omega_m) = 0\,. \qquad (5.54)$$

Note that the proof of this theorem makes use of the fact that m vanishes on the rotation axis. This is used twice in order to argue that the vanishing of the derivative of a function implies that the latter vanishes itself: First, the argument is applied to conclude that $d[F(k,m)] = 0$ and $d[(*F)(k,m)] = 0$ imply Ricci–circularity. Since the Ricci–circularity conditions are the *derivatives* of the Frobenius conditions (see eq. (2.78)), the argument is used again to conclude that the latter hold as well. As we already emphasized, the corresponding problem requires a more involved investigation if only *one* Killing field is present. In fact, the *staticity* theorem (Lichnerowicz 1955, Hawking and Ellis 1973, Carter 1973a, 1987) was proven only recently for the electrovac case by Sudarsky and Wald (1992). We shall turn our attention to this problem in section 8.2.

We conclude this section by evaluating the algebraic identities (5.19) and (5.20) for the case where k and m fulfil the integrability conditions. The metric of a circular spacetime assumes the form

$^{(4)}g = \sigma + X^{-1}\gamma$ (see chapter 3), implying that the 4–dimensional Hodge dual of $(\Omega \wedge E_m)$ becomes (using $k \wedge m = -\rho^2(dt \wedge d\varphi)$)

$$* (\Omega \wedge E_m) = -\rho^2 * (dt \wedge d\varphi \wedge E_m) = \rho *^{(\gamma)} E_m. \qquad (5.55)$$

Using the fact that $F(k,m)$ and $(*F)(k,m)$ vanish, the identities (5.19) and (5.20) for B_k and E_k in terms of B_m and E_m now become

$$B_k = A B_m - \frac{\rho}{X} *^{(\gamma)} E_m, \qquad (5.56)$$

$$E_k = A E_m + \frac{\rho}{X} *^{(\gamma)} B_m, \qquad (5.57)$$

where, as in chapter 3, $A = W/X$ and $\rho = \sqrt{-\sigma}$. In terms of the complex potentials Λ_k and Λ_m, defined by $d\Lambda_k = i_k(F + i * F) = -E_k + iB_k$ and $d\Lambda_m = i_m(F + i * F) = -E_m + iB_m$, respectively (see eq. (5.28)), we now have

$$d\Lambda_k = (A + i \frac{\rho}{X} *^{(\gamma)}) d\Lambda_m. \qquad (5.58)$$

This relation provides a helpful tool in establishing the connection between conjugate solutions of the Ernst equations (see section 5.6).

It is worth noticing that circularity also implies that the vector fields E_k, B_k, E_m and B_m (associated with the corresponding 1–forms) are perpendicular to the orbit manifold (Σ, σ). (For instance, $(k|E_k) = 0$ by definition and $(m|E_k) = 0$ since $F(k,m)$ vanishes in the circular case.) This also implies that the Poynting vectors (1–forms) assigned to k and m,

$$S_k = *(k \wedge E_k \wedge B_k), \qquad S_m = *(m \wedge E_m \wedge B_m), \qquad (5.59)$$

are tangent to (Σ, σ). Hence, in a circular, electrovac spacetime one has

$$(E_k)^a = (B_k)^a = (E_m)^a = (B_m)^a = 0, \qquad a = 0, 1, \quad (5.60)$$
$$(S_k)^i = (S_m)^i = 0, \qquad i = 2, 3. \quad (5.61)$$

(Since the metric is circular, the vanishing of contravariant and covariant components is equivalent.)

5.5 The circular Einstein–Maxwell equations

Now that the circularity theorem for the Einstein–Maxwell system is established, we can apply corollary 3.1: The metric can be written in the Papapetrou form (3.31) and the terms in eqs. (3.32)-(3.37) containing components of the Ricci tensor can be evaluated as follows: First, we have

$$\text{tr}_\sigma \boldsymbol{R} = 0, \tag{5.62}$$

which is due to the algebraic identity (5.22) and the consequence $F(k,m) = (*F)(k,m) = 0$ of the symmetry conditions and the Maxwell equations:

$$\text{tr}_\sigma \boldsymbol{R} = \frac{8\pi}{VX + W^2}[-XT(k,k) + VT(m,m) + 2WT(k,m)] = 0.$$

Secondly, eq. (5.9) for the stress–energy tensor (with $K = m$, $N = X$) together with $(S_m)^i = 0$ (see eq. (5.61)) yields

$$\begin{aligned} R_{ij} &= \frac{2}{X}[(E_m)_i(E_m)_j + (B_m)_i(B_m)_j] \\ &- \frac{1}{X}g_{ij}[(E_m|E_m)^{(g)} + (B_m|B_m)^{(g)}], \end{aligned} \tag{5.63}$$

for the nonvanishing components of the Ricci tensor. Using eq. (5.12) for $K = m$ and the 2–dimensional conformal Riemannian metric $\gamma = Xg$ (see eq. (3.22)), we thus have

$$R_{ij} + \frac{1}{X^2}\gamma_{ij}R(m,m) = \frac{1}{X}[\partial_i\Lambda\partial_j\bar{\Lambda} + \partial_j\Lambda\,\partial_i\bar{\Lambda}]. \tag{5.64}$$

In the remainder of this chapter we use Λ to denote the complex potential which is obtained from the electric and magnetic components of F with respect to the *axial* Killing field m,

$$d\Lambda \equiv d\Lambda_m = -E_m + iB_m. \tag{5.65}$$

Finally, eq. (3.32) is replaced by the two Ernst equations (5.41) for E and Λ, where the 4–dimensional Laplacian reduces to

$$\Delta\text{E} = \Delta^{(g)}\text{E} + \frac{1}{\rho}(d\rho|d\text{E})^{(g)} = -\frac{X}{\rho}d^{\dagger(\gamma)}(\rho\,d\text{E}), \tag{5.66}$$

where E denotes the *electrovac* Ernst potential, defined in eq. (5.38).

In combination with the above results, corollary 3.1 implies the following corollary:

Corollary 5.8 *Let* $(M, {}^{(4)}g)$ *denote an asymptotically flat, stationary and axisymmetric electrovac spacetime with Killing fields* k, m *and stationary and axisymmetric electromagnetic field* F. *Then* $(M, {}^{(4)}g)$ *is circular with metric* ${}^{(4)}g = \sigma + g$,

$$ {}^{(4)}g = -\frac{\rho^2}{X} dt^2 + X (d\varphi + A\, dt)^2 + \frac{1}{X}\gamma. \tag{5.67}$$

The Einstein–Maxwell equations reduce to the following set of equations on the 2–dimensional Riemannian manifold (Γ, γ):

$$\Delta^{(\gamma)}\rho = 0, \tag{5.68}$$

$$\frac{1}{\rho}\, dA = *^{(\gamma)}\Big(\frac{2\omega}{X^2}\Big), \tag{5.69}$$

$$\frac{1}{\rho}\, d^{\dagger(\gamma)}(\rho\, d\Lambda) = \frac{1}{X}\,(d\Lambda|d\mathrm{E} + 2\overline{\Lambda}d\Lambda)^{(\gamma)}, \tag{5.70}$$

$$\frac{1}{\rho}\, d^{\dagger(\gamma)}(\rho\, d\mathrm{E}) = \frac{1}{X}\,(d\mathrm{E}|d\mathrm{E} + 2\overline{\Lambda}d\Lambda)^{(\gamma)}, \tag{5.71}$$

$$\kappa^{(\gamma)}\gamma_{ij} - \frac{1}{\rho}\nabla_i^{(\gamma)}\nabla_j^{(\gamma)}\rho = \frac{\partial_i\Lambda\partial_j\overline{\Lambda} + c.c.}{X}$$

$$ + \frac{(\partial_i\mathrm{E} + 2\overline{\Lambda}\partial_i\Lambda)\,\overline{(\partial_j\mathrm{E} + 2\overline{\Lambda}\partial_j\Lambda)} + c.c.}{4\,X^2}, \tag{5.72}$$

where $d^{\dagger(\gamma)}$, $\Delta^{(\gamma)}$, $*^{(\gamma)}$ *and* $\kappa^{(\gamma)}$ *denote the co–derivative, the Laplacian, the Hodge dual and the Gauss curvature with respect to the metric* γ.
The complex Ernst potential E *is defined by*

$$\mathrm{E} = -(X + \Lambda\overline{\Lambda}) + iY, \tag{5.73}$$

where the complex electromagnetic potential Λ *is given by*

$$d\Lambda = i_m (F + i * F). \tag{5.74}$$

The real potential Y *is related to the twist* ω *(assigned to* m) *and* Λ *by*

$$dY = 2\omega - i\,(\Lambda\overline{d\Lambda} - \overline{\Lambda}d\Lambda). \tag{5.75}$$

As in the vacuum case, the function ρ is harmonic with respect to the 2–dimensional Laplacian of the orthogonal manifold (Γ, γ). Again, this enables one to introduce Weyl coordinates ρ and z, which yields the following generalization of proposition 4.1:

Proposition 5.9 *Let $(M,^{(4)}g)$ and F be as in corollary 5.8. Then the metric can be parametrized in terms of three functions X, A and h,*

$$^{(4)}g = -\frac{\rho^2}{X}\,dt^2 + X\,(d\varphi + A\,dt)^2 + \frac{1}{X}\,e^{2h}\,(d\rho^2 + dz^2)\,, \quad (5.76)$$

which are obtained from two complex potentials, Λ and E, by quadrature:

$$\frac{1}{\rho}\partial_\rho A = \frac{1}{X^2}\left[\partial_z Y + i\left(\Lambda\partial_z\overline{\Lambda} - \overline{\Lambda}\partial_z\Lambda\right)\right],$$

$$\frac{1}{\rho}\partial_z A = -\frac{1}{X^2}\left[\partial_\rho Y + i\left(\Lambda\partial_\rho\overline{\Lambda} - \overline{\Lambda}\partial_\rho\Lambda\right)\right], \quad (5.77)$$

$$\frac{1}{\rho}\partial_\rho h = \frac{1}{4X^2}\left[(\partial_\rho E + 2\overline{\Lambda}\partial_\rho\Lambda)\,\overline{(\partial_\rho E + 2\overline{\Lambda}\partial_\rho\Lambda)}\right]$$

$$+ \frac{1}{X}\left[\partial_\rho\Lambda\,\partial_\rho\overline{\Lambda}\right] - (\rho \leftrightarrow z)\,,$$

$$\frac{1}{\rho}\partial_z h = \frac{1}{4X^2}\left[(\partial_\rho E + 2\overline{\Lambda}\partial_\rho\Lambda)\,\overline{(\partial_z E + 2\overline{\Lambda}\partial_z\Lambda)}\right]$$

$$+ \frac{1}{X}\left[\partial_\rho\Lambda\,\partial_z\overline{\Lambda}\right] + (\rho \leftrightarrow z)\,, \quad (5.78)$$

where $-X = Re(E) + \Lambda\overline{\Lambda}$ and $Y = Im(E)$. The potentials E and Λ are subject to the Ernst equations,

$$\frac{1}{\rho}\underline{\nabla}(\rho\underline{\nabla}E) + \frac{(\underline{\nabla}E|\underline{\nabla}E + 2\overline{\Lambda}\underline{\nabla}\Lambda)^{(\delta)}}{X} = 0\,, \quad (5.79)$$

$$\frac{1}{\rho}\underline{\nabla}(\rho\underline{\nabla}\Lambda) + \frac{(\underline{\nabla}\Lambda|\underline{\nabla}E + 2\overline{\Lambda}\underline{\nabla}\Lambda)^{(\delta)}}{X} = 0\,, \quad (5.80)$$

which are also the Euler–Lagrange equations for the effective action

$$S_{em} = 4\int_\Gamma \kappa^{(\gamma)}\,\rho\,\eta^{(\gamma)} = 4\int_\Gamma \kappa^{(\gamma)}\,\sqrt{\gamma}\,\rho\,d\rho dz\,, \quad (5.81)$$

where $\kappa^{(\gamma)}$ denotes the Gauss curvature of the 2–dimensional metric γ and $\underline{\nabla} = (\partial_\rho, \partial_z)$.

Proof It remains to show that S_{em} is the effective action for the Ernst equations (5.79), (5.80). Taking the trace of equation (5.72) with respect to γ and using $\gamma^{ij}\nabla_i^{(\gamma)}\nabla_j^{(\gamma)}\rho = 0$ yields

$$\sqrt{\gamma}\,\kappa^{(\gamma)} = \frac{(d\mathrm{E} + 2\overline{\Lambda}d\Lambda\,|\,\overline{d\mathrm{E} + 2\overline{\Lambda}d\Lambda})^{(\delta)}}{4\,X^2} + \frac{(d\Lambda\,|\,\overline{d\Lambda})^{(\delta)}}{X}\,, \qquad (5.82)$$

where $X = X(\mathrm{E}, \overline{\mathrm{E}}, \Lambda, \overline{\Lambda}) = (-1/2)(\mathrm{E} + \overline{\mathrm{E}} + 2\Lambda\overline{\Lambda})$, and where we have also used $\sqrt{\gamma}(\cdot\,|\,\cdot)^{(\gamma)} = (\cdot\,|\,\cdot)^{(\delta)}$.

We already established the variational principle for the general action (5.42). Evaluating this expression by using the metric (5.76) and taking into account that both E and Λ depend only on ρ and z, we find

$$S[\mathrm{E}, \overline{\mathrm{E}}, \Lambda, \overline{\Lambda}] = \int_M \left[\frac{|d\mathrm{E} + 2\overline{\Lambda}d\Lambda|^2}{X^2} + 4\frac{|d\Lambda|^2}{X} \right] \sqrt{\gamma}\sqrt{|\sigma|}\,dx^0 dx^1 d\rho dz$$

$$= 4 \int_\Gamma \kappa^{(\gamma)}\sqrt{\gamma}\,\rho\,d\rho dz \int_\Sigma dx^0 dx^1 = \mathrm{vol}(\Sigma)\,S_{\mathrm{em}}\,,$$

where $|d\Lambda|^2 \equiv (d\Lambda\,|\,\overline{d\Lambda})^{(\gamma)}$. Hence, the effective Lagrangian with Euler–Lagrange equations (5.79) and (5.80) is given by $S_{\mathrm{em}} = \int_\Gamma \mathcal{L}_{\mathrm{em}}d\rho dz$, where

$$\mathcal{L}_{\mathrm{em}} = \rho \left[\frac{|\nabla\mathrm{E} + 2\overline{\Lambda}\nabla\Lambda|^2}{X^2} + 4\frac{|\nabla\Lambda|^2}{X} \right] \qquad (5.83)$$

and $|\nabla\Lambda|^2 \equiv (\nabla\Lambda\,|\,\nabla\overline{\Lambda})^{(\delta)}$. $\qquad\square$

Note that the Ernst equations (5.79) and (5.80) are the integrability conditions for the differential equations (5.78). This generalizes the corresponding result established in section 3.4 for the vacuum case.

5.6 The Kerr–Newman solution

The last section of this chapter is devoted to the derivation of the Kerr–Newman solution, describing the gravitational and electromagnetic fields of a rotating charged black hole (Newman et al. 1965). Introducing the new potentials $\epsilon = (1 + \mathrm{E})/(1 - \mathrm{E})$ and $\lambda = 2\Lambda/(1 - \mathrm{E})$, one finds that the Ernst equations (5.79), (5.80) admit the simple solution $\epsilon = px + iqy$, $\lambda = \lambda_0$, where $p, q \in \mathbb{R}$ and $\lambda_0 \in \mathbb{C}$ are constants. The equations for the metric functions A and h in terms of the prolate spheroidal coordinates x, y are almost the same as in the vacuum case and can be integrated without further difficulties. Again, the final step in the derivation of the solution is the construction of the conjugate metric. We shall see that the conjugate electric and magnetic fields $\hat{E}_k$ and $\hat{B}_k$ are identical to the fields E_m and B_m of the direct solution.

With respect to prolate spheroidal coordinates x and y (see eqs. (4.14), (4.15)), the Ernst equations (5.79) and (5.80) for the new potentials

$$\epsilon = \frac{1 + E}{1 - E}, \qquad \lambda = \frac{2\Lambda}{1 - E} \tag{5.84}$$

become (see eq. (4.21))

$$\Delta^{(\tilde{\gamma})}\epsilon + \frac{2}{1 - \epsilon\bar{\epsilon} - \lambda\bar{\lambda}}(\nabla\epsilon|\bar{\epsilon}\nabla\epsilon + \bar{\lambda}\nabla\lambda)^{(\tilde{\gamma})} = 0, \tag{5.85}$$

$$\Delta^{(\tilde{\gamma})}\lambda + \frac{2}{1 - \epsilon\bar{\epsilon} - \lambda\bar{\lambda}}(\nabla\lambda|\bar{\epsilon}\nabla\epsilon + \bar{\lambda}\nabla\lambda)^{(\tilde{\gamma})} = 0, \tag{5.86}$$

where the metric $\tilde{\gamma}$ is conformally related to γ by

$$\tilde{\gamma} = [e^{2h}\mu^2(x^2 - y^2)]^{-1}\gamma = (\frac{dx^2}{x^2 - 1} + \frac{dy^2}{1 - y^2}). \tag{5.87}$$

In terms of ϵ and λ, the effective action (5.83) assumes the symmetric form

$$S_{em} = 4\mu \int_\Gamma \frac{(1 - \lambda\bar{\lambda})(\nabla\epsilon|\nabla\bar{\epsilon})^{(\tilde{\gamma})} + \epsilon\bar{\lambda}(\nabla\bar{\epsilon}|\nabla\lambda)^{(\tilde{\gamma})} + (\epsilon \leftrightarrow \lambda)}{(1 - \epsilon\bar{\epsilon} - \lambda\bar{\lambda})^2}dxdy, \tag{5.88}$$

which reduces to the vacuum action (4.25) if λ vanishes.

Clearly $\lambda = $ constant fulfils the second Ernst equation (5.86). In addition, the first Ernst equation (5.85) (and the above action) reduce for constant values of λ to the corresponding vacuum expressions, up to an extra constant in the denominators. Hence, the ansatz

$$\epsilon = px + iqy, \qquad \lambda = \lambda_0, \tag{5.89}$$

solves both Ernst equations, provided that the real constants p and q are subject to the condition

$$p^2 + q^2 = 1 - \lambda_0\bar{\lambda_0}. \tag{5.90}$$

Note that in terms of the original potentials, E and Λ, the ansatz $\lambda = $ constant corresponds to the linear relation

$$\Lambda = \frac{\lambda_0}{2}(1 - E). \tag{5.91}$$

In terms of ϵ and λ_0, the functions X and Y and the twist 1–form ω become, respectively

$$X = \frac{1 - \epsilon\bar{\epsilon} - \lambda_0\overline{\lambda_0}}{|1 + \epsilon|^2}, \qquad Y = i\,\frac{\bar{\epsilon} - \epsilon}{|1 + \epsilon|^2}, \tag{5.92}$$

$$\omega = \frac{X^2}{2}\,i\,\frac{(\epsilon + 1)(\epsilon + 1 - \lambda_0\overline{\lambda_0})\,d\bar{\epsilon} - \text{c.c.}}{(1 - \epsilon\bar{\epsilon} - \lambda_0\overline{\lambda_0})^2}, \tag{5.93}$$

since $\omega = (1/2)(dY + i(\Lambda\overline{d\Lambda} - \overline{\Lambda}d\Lambda))$. The following equations for the partial derivatives of the metric functions A and h with respect to x and y are obtained in a similar manner to the vacuum case (see eqs. (4.22) and (4.23)):

$$\partial_x A = \mu\,(1 - y^2)\frac{2\omega_y}{X^2}, \qquad \partial_y A = -\mu\,(x^2 - 1)\frac{2\omega_x}{X^2}, \tag{5.94}$$

$$\frac{\partial_x h}{(1 - \lambda_0\overline{\lambda_0})} = \frac{1 - y^2}{x^2 - y^2}x\left[\frac{(x^2 - 1)\partial_x\epsilon\partial_x\bar{\epsilon} - (1 - y^2)\partial_y\epsilon\partial_y\bar{\epsilon}}{(1 - \epsilon\bar{\epsilon} - \lambda_0\overline{\lambda_0})^2} - \cdots\right],$$

$$\frac{\partial_y h}{(1 - \lambda_0\overline{\lambda_0})} = \frac{x^2 - 1}{x^2 - y^2}y\left[\frac{(x^2 - 1)\partial_x\epsilon\partial_x\bar{\epsilon} - (1 - y^2)\partial_y\epsilon\partial_y\bar{\epsilon}}{(1 - \epsilon\bar{\epsilon} - \lambda_0\overline{\lambda_0})^2} - \cdots\right],$$
$$\tag{5.95}$$

with $2\omega/X^2$ according to equation (5.93). The dots in the above equations stand for terms proportional to $(\partial_x\epsilon\partial_y\bar{\epsilon} + \partial_y\epsilon\partial_x\bar{\epsilon})$, which do not contribute for $\epsilon = px + iqy$. Inserting the solution (5.89) finally gives the slightly modified equations (see eqs. (4.29), (4.30))

$$\partial_x A = 2q\mu\,(1 - y^2)\frac{(1 + px)(1 + px - \lambda_0\overline{\lambda_0}) - (qy)^2}{[1 - (px)^2 - (qy)^2 - \lambda_0\overline{\lambda_0}]^2},$$

$$\partial_y A = 2p\mu\,(x^2 - 1)\frac{(1 + px - \lambda_0\overline{\lambda_0}/2)\,2qy}{[1 - (px)^2 - (qy)^2 - \lambda_0\overline{\lambda_0}]^2}, \tag{5.96}$$

$$\partial_x h = -\frac{(1 - \lambda_0\overline{\lambda_0})\,x\,(1 - y^2)}{(x^2 - y^2)[1 - (px)^2 - (qy)^2 - \lambda_0\overline{\lambda_0}]},$$

$$\partial_y h = -\frac{(1 - \lambda_0\overline{\lambda_0})\,y\,(x^2 - 1)}{(x^2 - y^2)[1 - (px)^2 - (qy)^2 - \lambda_0\overline{\lambda_0}]}. \tag{5.97}$$

Integration of these equations yields the solutions

$$A = 2\mu\,(q/p)\frac{(1 - y^2)(1 + px - \lambda_0\overline{\lambda_0}/2)}{1 - (px)^2 - (qy)^2 - \lambda_0\overline{\lambda_0}}, \tag{5.98}$$

$$e^{2h} = \frac{1 - (px)^2 - (qy)^2 - \lambda_0\overline{\lambda_0}}{p^2(x^2 - y^2)}. \qquad (5.99)$$

For the same reasons as in the vacuum case, the above solution is physically not acceptable. In order to obtain the correct solution, one must again consider the conjugate metric functions. These are obtained from the transformation (4.10), using the algebraic identity

$$\frac{4q^2(1-y^2)(1+px-\lambda_0\overline{\lambda_0}/2)^2 - p^2(x^2-1)\left[(1+px)^2 + (qy)^2\right]^2}{1 - (px)^2 - (qy)^2 - \lambda_0\overline{\lambda_0}}$$

$$= \left[(1+px)^2 + q^2\right]^2 - q^2p^2(x^2-1)(1-y^2), \qquad (5.100)$$

which generalizes the corresponding vacuum formula (4.33). Using $\hat{V} = -X$ and $\hat{W} = \hat{A}\hat{X} = -AX$ eventually yields the desired result:

$$\hat{X} = \frac{\mu^2}{p^2}(1-y^2)\frac{[(1+px)^2 + q^2]^2 - p^2q^2(x^2-1)(1-y^2)}{(1+px)^2 + (qy)^2}, \qquad (5.101)$$

$$\hat{A} = -2\frac{qp}{\mu}\frac{(1+px-\lambda_0\overline{\lambda_0}/2)}{[(1+px)^2 + q^2]^2 - p^2q^2(x^2-1)(1-y^2)}, \qquad (5.102)$$

$$\hat{V} = -\frac{1 - (px)^2 - (qy)^2 - \lambda_0\overline{\lambda_0}}{(1+px)^2 + (qy)^2}, \qquad (5.103)$$

$$\hat{W} = -2\frac{\mu q}{p}\frac{(1-y^2)(1+px-\lambda_0\overline{\lambda_0}/2)}{(1+px)^2 + (qy)^2}, \qquad (5.104)$$

$$e^{2\hat{h}} = \hat{X}\frac{(1+px)^2 + (qy)^2}{p^2(x^2-y^2)}, \qquad (5.105)$$

$$\hat{Y} = 2\frac{q\mu^2}{p^2}y[(3-y^2)+(1-y^2)\frac{q^2(1-y^2) - \lambda_0\overline{\lambda_0}(1+px)}{(1+px)^2 + (qy)^2}]. \qquad (5.106)$$

The electromagnetic potential $\Lambda \equiv \Lambda_m$ corresponding to the *unconjugated* solution, becomes

$$\Lambda_m = \frac{\lambda_0}{1+\epsilon} = \frac{\lambda_0(1 + px - iqy)}{(1+px)^2 + (qy)^2}. \qquad (5.107)$$

As we shall see in a moment, it is not necessary to compute the conjugate potential to obtain the correct electric and magnetic fields. (The reason for this lies in the fact that Λ_m is the unconjugated potential for the electric and magnetic components of F with respect to the *axial* Killing field. We are, however, interested in the conjugate electric and magnetic components of F with respect to the *stationary* Killing field, that is, in $\hat{E}_k$ and $\hat{B}_k$.)

Let us again rewrite the above expressions in terms of Boyer–Lindquist coordinates,

$$r = m\,(1 + px)\,, \quad \cos\vartheta = y\,. \tag{5.108}$$

In addition to the parameters m and a (see eq. (4.42)), we introduce the real parameter Q. Restricting ourselves to real values of λ_0, we define

$$m = \mu\,p^{-1}\,, \quad a = \mu\,q\,p^{-1}\,, \quad Q = \mu\,\lambda_0\,p^{-1}\,, \tag{5.109}$$

where $p^2 + q^2 = 1 - \lambda_0\overline{\lambda_0}$ implies

$$\mu^2 = m^2 - a^2 - Q^2\,. \tag{5.110}$$

(A nonvanishing imaginary part of λ_0 corresponds to a magnetic charge.) Defining Δ and Ξ by (see eq. (4.43))

$$\Delta = r^2 - 2mr + a^2 + Q^2\,, \quad \Xi = r^2 + a^2\cos^2\vartheta\,, \tag{5.111}$$

we see that the relations (4.44)-(4.46) remain unchanged. Instead of the identities (4.47) and (4.48) we now have

$$(px)^2 + (qy)^2 + \lambda_0\overline{\lambda_0} - 1 = m^{-2}\,(\Delta - a^2\sin^2\vartheta)\,, \tag{5.112}$$

$$1 + px - \lambda_0\overline{\lambda_0} = \frac{r}{m} - \frac{Q^2}{2m^2} = \frac{(r^2 + a^2) - \Delta}{2\,m^2}\,. \tag{5.113}$$

By virtue of these formulae, the metric functions $\hat{X}$, $\hat{W}$, $\hat{V}$ and $\hat{h}$ can be expressed in terms of Δ and Ξ with the same relations as in the vacuum case (see eqs. (4.49)-(4.52)). Hence, the Kerr–Newman metric becomes

$$^{(4)}g = \frac{1}{\Xi}[-(\Delta - a^2\sin^2\vartheta)\,dt^2 + 2a\,\sin^2\vartheta\,(\Delta - (r^2 + a^2))\,dt d\varphi$$

$$+ \sin^2\vartheta\,((r^2 + a^2)^2 - \Delta a^2\sin^2\vartheta)\,d\varphi^2] + \Xi[\frac{1}{\Delta}\,dr^2 + d\vartheta^2]\,, \tag{5.114}$$

with Δ and Ξ according to eq. (5.111). In addition, the imaginary part $\hat{Y}$ of the conjugate Ernst potential and the *unconjugated* complex potential Λ_m become, respectively

$$\hat{Y} = 2am\,\cos\vartheta[(3 - \cos^2\vartheta) + \sin^2\vartheta\,\frac{a^2\sin^2\vartheta - (r/m)Q^2}{\Xi}]\,, \tag{5.115}$$

$$\Lambda_m = Q\,\frac{r - i\,a\,\cos\vartheta}{\Xi}\,. \tag{5.116}$$

The Kerr–Newman metric describes a 3–parameter family of asymptotically flat, stationary and axisymmetric solutions to the

Einstein–Maxwell equations. For $a = 0$, it reduces to the static, spherically symmetric Reissner–Nordström solution, whereas it coincides with the vacuum Kerr solution if Q vanishes.

In order to complete the derivation of the Kerr–Newman solution, it remains to compute the conjugate electric and magnetic fields, $\hat{E}_k$ and $\hat{B}_k$. At first glance this might seem to be laborious, since our derivation was based on the fields E_m and B_m. However, as we shall prove at the end of this section, the unconjugated electric and magnetic components of F with respect to m are identical to the conjugate components with respect to k:

$$\hat{E}_k = E_m, \qquad \hat{B}_k = B_m. \qquad (5.117)$$

Hence, using $d\Lambda \equiv d\Lambda_m = -E_m + iB_m$, the conjugate electric and magnetic fields become

$$\hat{E}_k = \frac{Q}{\Xi^2} \left[(r^2 - a^2 \cos^2\vartheta)\, dr - 2ra^2 \cos\vartheta \, \sin\vartheta \, d\vartheta \right], \qquad (5.118)$$

$$\hat{B}_k = \frac{Q}{\Xi^2} \left[(r^2 - a^2 \cos^2\vartheta) \sin\vartheta \, d\vartheta + 2r \cos\vartheta \, dr \right]. \qquad (5.119)$$

Taking advantage of eq. (5.7) (with $K = k$ and $N = -V$), this eventually yields the electromagnetic 2–form $\hat{F}$,

$$\hat{F} = \frac{Q}{\Xi^2} (r^2 - a^2 \cos^2\vartheta) \left[dr \wedge \{dt - a \sin^2\vartheta \, d\varphi\} \right]$$
$$+ \frac{Q}{\Xi^2} 2r\, a \, \cos\vartheta \, \sin\vartheta [d\vartheta \wedge \{ (r^2 + a^2)\, d\varphi - a \, dt \}], \qquad (5.120)$$

and the associated vector potential $\hat{A}$,

$$\hat{A} = \frac{Q}{\Xi} r \{ dt - a \sin^2\vartheta \, d\varphi \}. \qquad (5.121)$$

The asymptotic behavior of the electric field suggests that Q is the total charge of the system. In fact, we have the following corollary:

Corollary 5.10 *The quantities m, a and Q parametrizing the Kerr–Newman metric (5.114) are identical to the total mass, M, the angular momentum per unit mass, J/M, and the total electric charge, respectively.*

Proof The asymptotic behavior of $*dk$ and $*dm$ remains the same as for the vacuum solution (4.54), which implies that m and ma

are equal to the total mass M and the total angular momentum J, respectively. Using the expressions (4.55) and (4.56), we obtain

$$
\begin{aligned}
(*\hat{F})_{\vartheta\varphi} &= \frac{Q}{\Xi^2}\,(r^2 - a^2\cos^2\vartheta)\,[\,(*(dr \wedge dt))_{\vartheta\varphi} \\
&\quad - a\,\sin^2\vartheta\,(*(dr \wedge d\varphi))_{\vartheta\varphi}\,] \\
&= \frac{Q}{r^2}\,[\,r^2\sin\vartheta + \mathcal{O}(r)\,],
\end{aligned}
\tag{5.122}
$$

which yields

$$
\frac{1}{4\pi}\int_{S_2^\infty} *\hat{F} = \frac{Q}{4\pi}\int_{S_2^\infty}\sin\vartheta\,d\vartheta d\varphi = Q .
\tag{5.123}
$$

This establishes that the parameter Q is the charge as seen by an asymptotic observer. $\qquad\square$

This completes our derivation of the Kerr–Newman metric. We refer the reader to the literature for a discussion of the highly nontrivial properties of the Kerr–Newman geometry. We conclude this chapter by giving the boundary conditions for an asymptotically flat, stationary and axisymmetric electrovac spacetime. We must also state the proof of the following proposition, which was applied in order to find the conjugate electric and magnetic fields:

Proposition 5.11 *Let* (X, A, h, Λ_m) *be a solution of the circular electrovac equations. Then the conjugate quantities* $(\hat{X}, \hat{A}, \hat{h}, \hat{\Lambda}_m)$ *solve the same equations, provided that* $\hat{X}$, $\hat{A}$ *and* $\hat{h}$ *are related to* X, A *and* h *according to eq. (4.10) and*

$$
- d\hat{\Lambda}_m = (A + i\,\frac{\rho}{X}\,*^{(\gamma)})\,d\Lambda_m .
\tag{5.124}
$$

This also implies

$$
\hat{E}_k = E_m , \qquad \hat{B}_k = B_m .
\tag{5.125}
$$

Proof Following Chandrasekhar (1983), we first demonstrate that Λ fulfils the equation

$$
d\left[\left(A + i\,\frac{\rho}{X}\,*^{(\gamma)}\right)d\Lambda_m\right] = 0 .
\tag{5.126}
$$

Substituting $dE + 2\bar{\Lambda}_m d\Lambda_m$ by $-dX + 2i\omega$ in eq. (5.70) and using

$$
\frac{1}{\rho}\,d^{\dagger(\gamma)}\left(\frac{\rho}{X}\,d\Lambda_m\right) = \frac{1}{\rho}\,\frac{d^{\dagger(\gamma)}(\rho\,d\Lambda_m)}{X} + \frac{(d\Lambda_m | dX)^{(\gamma)}}{X^2}
$$

and eq. (3.13), relating the dual of the twist–form to the derivative of A, we find

$$d *^{(\gamma)} \left(\frac{\rho}{X} d\Lambda_m \right) = - (d\Lambda_m | \frac{\rho}{X^2} 2i\omega) *^{(\gamma)} 1,$$

$$- d\Lambda_m \wedge *^{(\gamma)} \left(\frac{\rho}{X^2} 2i\omega \right) = -i \, d\Lambda_m \wedge dA = i \, d(A \, d\Lambda_m),$$

which proves eq. (5.126). (Here we have used $(*^{(\gamma)})^2 = (-1)^p$ and $d^{\dagger(\gamma)} = - *^{(\gamma)} d*^{(\gamma)}$; see eqs. (1.14) and (1.17).)

We now show that $d\hat{\Lambda}_m$ and $d\Lambda_m$ *are* indeed conjugate quantities. Inverting the assertion (5.124) we find

$$- A \, d\hat{\Lambda}_m + i \frac{\rho}{X} *^{(\gamma)} d\hat{\Lambda}_m = (A^2 - \frac{\rho^2}{X^2}) \, d\Lambda_m \,,$$

or, equivalently, using the transformations $\hat{X} = X(A^2 - \rho^2/X^2)$ and $\hat{A} = -A(A^2 - \rho^2/X^2)^{-1}$,

$$d\Lambda_m = (\hat{A} + i \frac{\rho}{\hat{X}} *^{(\gamma)}) \, d\hat{\Lambda}_m \,. \tag{5.127}$$

The integrability condition $dd\Lambda_m = 0$ for this expression yields the same equation for $d\hat{\Lambda}_m$ in terms of $\hat{X}$ and $\hat{A}$ as eq. (5.126) for $d\Lambda_m$ in terms of X and A. Hence, $d\hat{\Lambda}_m$ and $d\Lambda_m$ are conjugate quantities. We can now apply the conjugate version of the identity (5.58) and eq. (5.127) to write

$$-\hat{E}_k + i\hat{B}_k = d\hat{\Lambda}_k = (\hat{A} + i \frac{\rho}{\hat{X}} *^{(\gamma)}) \, d\hat{\Lambda}_m = d\Lambda_m = -E_m + i \, B_m \,.$$

$$\square$$

For the sake of completeness, we also give the explicit expression for $\hat{\Lambda}_m$. This is obtained from integrating equation (5.124) and using the expressions for X, A and ρ. A straightforward but tedious computation yields

$$\begin{aligned}
\hat{\Lambda}_m &= Q \left[-q(1 - y^2) \frac{1 + px}{(1 + px)^2 + (qy)^2} + iy \frac{(1 + px)^2 + q^2}{(1 + px)^2 + (qy)^2} \right] \\
&= Q \left[- \sin^2\vartheta \, \frac{ar}{\Xi} + i \cos\vartheta \, \frac{r^2 + a^2}{\Xi} \right] \,. \tag{5.128}
\end{aligned}$$

We finally note that the behavior of the Kerr–Newman potentials at infinity, the rotation axis, and the horizon is in accordance with the following, general conditions (see, e.g., Mazur 1984a, Carter 1985, Weinstein 1990):

(i) The asymptotic form (2.33) of a stationary and axisymmetric, asymptotically flat metric, together with the asymptotic conditions $\hat{\phi}_k = -(Q/r)[1 + \mathcal{O}(1/r)]$, $\hat{\psi}_k = -(Q/r^2)a\cos\vartheta[1 + \mathcal{O}(1/r)]$, yields

$$
\begin{aligned}
\hat{X} &= (1-y^2)\left[\mu^2 x^2 + \mathcal{O}(x)\right], \\
\hat{Y} &= 2Jy(3-y^2) + \mathcal{O}(x^{-1}), \\
\hat{\Lambda}_m &= -(1-y^2)\frac{QJ}{M}\left[(\mu x)^{-1} + \mathcal{O}(x^{-2})\right] \\
&\quad + iQ\left[y + \mathcal{O}(x^{-2})\right].
\end{aligned}
\tag{5.129}
$$

(ii) In the vicinity of the rotation axis, i.e., for $y \to \pm 1$, we need

$$
\begin{aligned}
\hat{X} &= \mathcal{O}(1-y^2), & (1-y^2)\frac{\partial_y \hat{X}}{\hat{X}} &= \mp 2 + \mathcal{O}(1-y^2), \\
\hat{Y} &= \pm 4J + \mathcal{O}(1-y^2), & \partial_x \hat{Y} &= \mathcal{O}((1-y^2)^2), \\
\hat{\Lambda}_m &= \pm iQ + \mathcal{O}(1-y^2), & \partial_x \hat{\Lambda}_m &= \mathcal{O}(1-y^2).
\end{aligned}
\tag{5.130}
$$

This also shows that both components of the twist are of $\mathcal{O}(1-y^2)$,

$$
\hat{\omega} = \frac{1}{2}\left[d\hat{Y} + i(\hat{\Lambda}_m d\overline{\hat{\Lambda}}_m - \overline{\hat{\Lambda}}_m d\hat{\Lambda}_m)\right] = \mathcal{O}(1-y^2),
\tag{5.131}
$$

since the terms which are not of $\mathcal{O}(1-y^2)$ cancel each other in the expressions for $d\hat{Y}$ and $i(\hat{\Lambda}_m d\overline{\hat{\Lambda}}_m - \overline{\hat{\Lambda}}_m d\hat{\Lambda}_m)$.

(iii) On the horizon, i.e., for $x = 1$, the potentials must behave regularly,

$$
\begin{aligned}
\hat{X} &= \mathcal{O}(1), & \hat{X}^{-1} &= \mathcal{O}(1), \\
\partial_x \hat{Y} &= \mathcal{O}(1), & \partial_y \hat{Y} &= \mathcal{O}(1), \\
\partial_x \hat{\Lambda}_m &= \mathcal{O}(1), & \partial_y \hat{\Lambda}_m &= \mathcal{O}(1).
\end{aligned}
\tag{5.132}
$$

6

Stationary black holes

In 1931 Chandrasekhar established an upper bound for the mass of a cold self–gravitating star in thermal equilibrium (Chandrasekhar 1931a, 1931b). This leads one to consider the ultimate fate of a star which, having radiated all its thermo–nuclear energy, still has a mass beyond the critical limit (a few solar masses). Once the nickel and iron core has been formed, there exists no possibility for any further nuclear reactions; the core must therefore undergo gravitational collapse. The collapse may cease by the time the core has reached nuclear densities, which leads to the formation of a neutron star, provided that the mass of the collapsing part lies below the critical value. If this is not the case, then nothing can prevent total gravitational collapse (Chandrasekhar 1939, Oppenheimer and Snyder 1939, Oppenheimer and Volkoff 1939), resulting in the formation of a black hole (Wheeler 1968; see Israel 1987 for a historical review).

Birkhoff's theorem (Birkhoff 1923), which states that a spherically symmetric spacetime is locally isometric to a part of the Schwarzschild–Kruskal metric (Kruskal 1960), yields a significant simplification in the discussion of the spherically symmetric collapse scenario (Harrison *et al.* 1965). However, in order to treat more general situations, one has to find the generic features of gravitational collapse in general relativity. This was achieved by Geroch, Hawking, Penrose and others in the late sixties and early seventies (Hawking and Penrose 1970; see also Hawking and Ellis 1973, Clarke 1975, 1993): The singularity theorems show that - in contrast to Newtonian gravity - deviations from spherical symmetry, internal pressure or rotation do not prevent the formation of a singularity. More precisely, the development of a singularity is inevitable once a marginally trapped surface has formed. Eventually, the apparent horizon (the boundary of the region where there are trapped surfaces) becomes the event horizon, representing the boundary of the black hole region of spacetime.

Numerous authors have contributed to the investigation of the global and local properties of black hole spacetimes. One of the most interesting predictions of the theory is the area increase theorem, which suggests a relationship between black hole physics and the second law of thermodynamics (see chapter 7). A further outcome of the general theory is the so–called strong rigidity theorem (Hawking 1972). Based mainly on the assumptions of asymptotic flatness and stationarity, it establishes the existence of a Killing field tangent to the event horizon. In particular, this implies that either the null–generators of the horizon coincide with the asymptotically timelike Killing field, or spacetime is stationary *and* axisymmetric. This subdivision of stationary black hole configurations into rotating and nonrotating ones must be considered the basis of the uniqueness theorems. In addition, the rigidity theorem can be applied to show that the surface gravity of a stationary black hole is uniform over the horizon, a fact which is known as the zeroth law of black hole physics (Bardeen *et al.* 1973, Carter 1973a).

This chapter is organized as follows: In the first section we recall the notions of black holes and event horizons. In the second section we discuss the strong rigidity theorem, which will underly the subsequent discussion. We shall, however, not go into the details of its proof, which lies beyond the scope of this text and is not required in order to understand the remainder of this chapter.

The weak rigidity theorem (Carter 1969, 1973b) is discussed in the third section. In contrast to the strong theorem, the weak version already assumes the existence of a second, axial Killing field (in addition to the stationary one). Applying a theorem due to Vishveshwara (1968), the Killing property of the horizon and the rigidity of its rotation are then established from the integrability conditions for the two Killing fields.

The fourth section is devoted to the general properties of Killing horizons. We introduce the surface gravity, κ, and give an efficient derivation of the fundamental fact that $R(K, K)$ vanishes on the horizon, where K is the horizon Killing field, i.e., the Killing field tangent to the null–generators of the horizon. Together with Einstein's equations and the dominant energy condition, this property will be used in the next chapter to show that the surface gravity is constant over the horizon.

The last section deals with the topology of the horizon. The derivation of the uniqueness result for nonrotating black holes is based on the assumption that each connected component $\mathcal{H}$ of the horizon "at time Σ" is a topological 2–sphere. Moreover, the uniqueness proof for rotating black holes requires a connected horizon with spherical topology. Here we present a brief outline of the arguments given in Hawking and Ellis (1973). Although strongly suggesting that each connected component of $\mathcal{H}$ has spherical topology, the proof does not seem to exclude the toroidal case in a conclusive way if the metric is not assumed to be analytic (see Galloway 1993, 1994). In fact, this problem was solved only recently by Chruściel and Wald (1994b) who - under fairly mild conditions - were able to show that the domain of outer communications (Geroch 1970a) is simply connected.

6.1 Basic definitions

In this section we briefly recall the basic concepts and definitions applying to black hole spacetimes. The notions are mainly adopted from Wald (1984, chap. 12). We refer the reader to this book and to the original literature (see, e.g., Hawking 1972, Hawking and Ellis 1973, Carter 1979 and refs. therein) for a detailed and more comprehensive presentation of the subject. Here we restrict ourselves to a descriptive introduction. We assume that the reader is familiar with the basic concepts of asymptotic flatness and conformal infinity and refer to Ashtekar (1980), Newman and Tod (1980) and Wald (1984, chaps. 7 and 11) for further reading.

A considerable body of work dealing with the theory of black holes - such as the classical uniqueness theorems - is restricted to stationary and asymptotically flat situations. The first requirement is based on the expectation that a black hole, representing the ultimate fate of a (sufficiently massive) collapsing star, settles down to a stationary configuration, once the excitations of the external fields are damped away by gravitational radiation. Unfortunately, the assumption of asymptotic flatness is, for instance, not appropriate when describing black holes in certain cosmological situations, such as in a recollapsing universe. However, it provides a convenient definition of a black hole in terms of its *event horizon* (Hawking 1973). This is a global concept, which requires the existence of conformal infinity:

Definition 6.1 *An asymptotically flat spacetime (M, g) is called strongly asymptotically predictable if there exists an open region $\tilde{V}$ in the conformal spacetime $(\tilde{M}, \tilde{g})$, with $\tilde{V} \supset \overline{M \cap J^-(\mathcal{J}^+)}$, such that $(\tilde{V}, \tilde{g})$ is globally hyperbolic.*

We recall that a spacetime is called globally hyperbolic if it admits a Cauchy surface, that is, a subset $\Sigma \in M$ which is intersected exactly once by every inextensible timelike trajectory of M.

The assumption that gravitational collapse results in a strongly asymptotically predictable spacetime (Hawking 1972) excludes the existence of naked singularities, that is, singularities which lie in the causal past of future null infinity $J^-(\mathcal{J}^+)$. This is, roughly speaking, the contents of the famous cosmic censor conjecture (Penrose 1969, 1979; see also Geroch and Horowitz 1979). More precisely, the conjecture states that the maximal Cauchy evolution of an asymptotically flat initial data set is an asymptotically flat, strongly asymptotically predictable spacetime, provided that the coupled Einstein and matter equations assume the form of a second order, quasi–linear, diagonal hyperbolic system and that the stress–energy tensor satisfies the dominant energy condition. (See Wald 1984 for a detailed discussion and alternative formulations.)

A spacetime contains a black hole if the past of all timelike curves in M having endpoints on future null infinity $\mathcal{J}^+$ is not the entire spacetime. More precisely, we have the following definition:

Definition 6.2 *The black hole region* B *of a strongly asymptotically predictable spacetime (M, g) is the part of M which is not contained in the causal past J^- of future null infinity $\mathcal{J}^+$, i.e., $B = M - J^-(\mathcal{J}^+)$.*

Definition 6.3 *The (future) event horizon H^+ is the boundary of the black hole region in M, $H^+ = M \cap \partial J^-(\mathcal{J}^+)$, i.e., the boundary of the region from which there is no escape to infinity for photons or particles.*

It also turns out to be convenient to introduce the concept of a black hole at a given "time Σ":

Definition 6.4 *Let Σ be a Cauchy surface for the globally hyperbolic region $\tilde{V}$ and let B be the black hole region. Then $\Sigma \cap B$ is called the black hole region at time Σ, and each connected component of $\Sigma \cap B$ is called a black hole at time Σ. The event horizon at time Σ is the 2-surface $\mathcal{H}^+ = \Sigma \cap H^+$.*

As already mentioned, the event horizon is a global concept since it is specified in terms of events at future infinity. Local criteria characterizing the presence of black holes include apparent horizons (Hawking 1972, Hawking and Ellis 1973; see also Hayward 1993) and trapped surfaces (Penrose 1965b). The future event horizon H^+ is an achronal 3-dimensional submanifold, each point of which lies in a future inextensible null geodesic segment which is entirely contained in H^+. We refer the reader to the literature mentioned above for more information on the evolution of achronal boundaries and the kinematics of null congruences. The theory is highlighted by Hawking's strong rigidity theorem, establishing the Killing property of the event horizon. Since this theorem represents the point of departure for a multitude of applications, we devote the next section to some comments concerning its content and domain of validity.

6.2 The strong rigidity theorem

The power of the strong rigidity theorem lies in the fact that it relates the concept of event horizons to the independently defined, and logically distinct, local notion of Killing horizons. In all that follows, the latter will be of fundamental importance. Here we adopt the definitions given by Chruściel (1994):

Definition 6.5 *Consider a Killing field K and the set of points on which K is null and not identically vanishing. Let $H_i[K]$ be a connected component of this set which is a null hypersurface. Any union $H[K] = \cup H_i[K]$ is called a Killing horizon.*

Roughly speaking, a Killing horizon is a null hypersurface to which a Killing field is normal. (Recall that the normal vector field to a null hypersurface is also tangent to it.) The connected components of the Killing horizon generated by the asymptotically timelike Killing field in Schwarzschild–Kruskal–Szekeres spacetime intersect in a 2-dimensional subset. This motivates the following definition:

Definition 6.6 *A bifurcate Killing horizon consists of a pair of Killing horizons (associated with the same Killing field) which intersect in a submanifold $\mathcal{F}$. The smooth, compact, embedded 2-surface $\mathcal{F}$ on which the Killing field must vanish is called the bifurcation surface.*

The existence of a nondegenerate Killing horizon does not automatically imply the existence of a bifurcation surface, although the converse statement is true. However, Rácz and Wald (1992) showed that one can find a local extension of a neighborhood of a *regular* Killing horizon such that the latter comprises a part of a bifurcate horizon. Moreover, the extension can be made *global* under certain weak conditions, which are automatically fulfilled in a static or circular spacetime (Rácz and Wald 1995). This strongly suggests that the only Killing horizons in spacetimes with matter satisfying the dominant energy condition are either of bifurcate type or degenerate (i.e., their surface gravity vanishes). Here we shall not go into further details of this problem and refer the reader to Kay and Wald (1991), Rácz and Wald (1992, 1995) and Chruściel (1994) for more information. It is, however, important to be aware of the fact that many derivations require the existence of a *bifurcate* Killing horizon.

As already mentioned, the logical basis of the black hole uniqueness theorems is the rigidity theorem. It guarantees that the isometry group of an asymptotically flat, stationary (but not static) analytic spacetime with complete Killing field is at least 2-dimensional. The proof proceeds in two steps:

Theorem 6.7 *(first part): The event horizon of a stationary black hole spacetime is a Killing horizon, provided that spacetime is analytic, the fundamental matter fields obey well behaved hyperbolic equations and the stress–energy tensor fulfils the weak energy condition.*

For the proof of this theorem we refer the reader to Hawking (1972) and Hawking and Ellis (1973). The reasoning given in Hawking (1972) requires the existence of both future and past event horizons, intersecting in the compact 2–surface $\mathcal{F} = \partial J^-(\mathcal{J}^+) \cap \partial J^+(\mathcal{J}^-)$: The first observation consists of the fact that the solution lying to the future of the bifurcation surface is determined by Cauchy data on the horizons $\partial J^-(\mathcal{J}^+)$ and $\partial J^+(\mathcal{J}^-)$. One then shows that the data are invariant under a continuous transformation which leaves $\mathcal{F}$ invariant and moves points on the one horizon toward $\mathcal{F}$ and points on the other horizon away from $\mathcal{F}$. Finally, one takes advantage of the uniqueness of the Cauchy development, in order to argue that the solution admits a Killing field which coincides with the null generator of the event horizon. As a consequence of this, one of the following alternatives must hold:

Theorem 6.8 *(second part): The horizon Killing field K either coincides with the stationary Killing field k, or the spacetime admits at least one axial Killing field.*

Definition 6.9 *A Killing field m which is tangent to the generators of an $SO(2)$ symmetry group with nonempty fixed point set is called an axial Killing field.*

Before we proceed, we would like to stress that the proof of the rigidity theorem is based on analyticity. This presents no restriction if the Killing field K is timelike, since the stationary field equations are elliptic in this case (Carter 1987). However, in regions where K becomes null or spacelike, the metric need not be analytic. In fact, even in situations without ergoregions, there is no justification for analyticity of the metric on the event horizon. As was pointed out by Chruściel (1994), the analyticity problem in the proof of the rigidity theorem is probably the most serious gap in the uniqueness theory of black holes (see also Chruściel 1991).

Regarding theorem 6.8, it should be mentioned that the classification of stationary, *asymptotically flat* spacetimes with at least two Killing fields is now understood quite well: This is due to the fact that, under some mild conditions, the Killing Lie algebra of spacetime is the Poincaré group if the additional Killing field is not axial (see Collinson 1970 and Chen 1981). We can therefore restrict ourselves to the two situations described in theorem 6.8: In the first case the black hole is said to be *nonrotating*. Provided that there exists no third Killing field - which would imply spherical symmetry - one has in the *rotating* situation

$$K = \ell = k + \Omega_H m. \tag{6.1}$$

Here and henceforth we use K to denote the generator of an arbitrary Killing horizon and ℓ for the generator of the Killing horizon of an axisymmetric black hole. (As usual, k and m denote the stationary and the axial Killing field, respectively.) The angular velocity of the horizon is denoted by Ω_H and the Killing field m is normalized such that its closed orbits have the period 2π. Note that, as a further consequence of asymptotic flatness, one has $[k, m] = 0$ (Carter 1970; see also theorem 2.15). Since the horizon is invariant under the symmetry groups generated by k and m, these Killing fields are tangent to $H[\ell]$. Using the fact that a tangential null vector is also orthogonal to $H[\ell]$, we obtain

$$(\ell|k) = 0, \quad (\ell|m) = 0 \quad \text{on } H[\ell]. \tag{6.2}$$

By virtue of eq. (6.1), this also implies

$$-\Omega_H = \frac{(k|m)}{(m|m)} = \frac{(k|k)}{(k|m)} \quad \text{on } H[\ell], \tag{6.3}$$

$$\mathcal{N} = (k|k)(m|m) - (k|m)^2 = 0 \quad \text{on } H[\ell], \tag{6.4}$$

where $\mathcal{N} \equiv -(k \wedge m | k \wedge m)$ (see eq. (2.79)). Equation (6.1) implies that both m and $\Omega_H m$ are Killing fields. Using the Killing equations we therefore have $m_\nu \nabla_\mu \Omega_H + m_\mu \nabla_\nu \Omega_H = 0$. Contracting this with $m^\mu m^\nu$ gives $L_m \Omega_H = 0$. Using this and contracting the above equation with m^μ yields $d\Omega_H = 0$ (if m is not null). The angular velocity assumes therefore a constant value over the horizon,

$$\Omega_H = \text{constant on } H[\ell]. \tag{6.5}$$

The simplicity of the above argument for the uniformity of Ω_H reflects the power of the strong rigidity theorem. If, on the other hand, spacetime is *assumed* to be stationary *and* axisymmetric, then the Killing property of the horizon and $\Omega_H = $ constant can be established by different means. We shall now present this so-called *weak* rigidity theorem, which is due to Carter (1969) and holds for two Killing fields which are subject to the integrability conditions.

6.3 The weak rigidity theorem

The strong rigidity theorem establishes the Killing property of a stationary event horizon, provided that the metric is analytic and the matter fields satisfy suitable hyperbolic equations and energy conditions. In this section we present a weaker version of this theorem. It establishes the Killing property of the horizon and the rigidity of its rotation for circular spacetimes (Carter 1969; see also DeFelice and Clarke 1990). We shall encounter a very similar situation when establishing the uniformity of the surface gravity (zeroth law) in the next chapter: Imposing integrability conditions considerably simplifies the proof of both the zeroth law and the rigidity theorem.

Before proving the weak rigidity theorem, we recall a theorem due to Vishveshwara (1968):

Theorem 6.10 *Let K be a Killing field and let $S[K]$ denote the surface $\{N \equiv (K|K) = 0\}$. Then $S[K]$ is a null surface (i.e., a Killing horizon, $S[K] = H[K]$) if and only if*

$$\omega = 0\,, \quad i_K\, dK \neq 0 \quad on \ S[K]. \qquad (6.6)$$

Proof Let $S[K]$ be a null surface, $S[K] = H[K]$. Then the 1-form dN is null on $H[K]$. In order to see that the twist vanishes on $H[K]$, we note that dN is orthogonal to K, since the action of K leaves $S[K]$ invariant,

$$(K|dN) = L_K N = (L_K g)(K, K) + 2\,(K|[K, K]) = 0\,.$$

Since two orthogonal null vectors are proportional, we also have $K \wedge dN = 0$. Together with the basic identity (2.10),

$$2 * (K \wedge \omega) = -\,(N\, dK + K \wedge dN)\,, \qquad (6.7)$$

and $N = 0$, this yields $\omega = 0$ on $H[K]$. Clearly $dN \neq 0$ implies $i_K dK \neq 0$, since $i_K dK = L_K K - d i_K K = -dN$.

Let $\omega = 0$, $i_K dK \neq 0$ on $S[K]$. Then we can apply the general identity (2.11) to deduce that

$$(dN|dN) = N(dK|dK) = 0 \quad \text{on } S[K].\qquad (6.8)$$

Since, by assumption, $dN \neq 0$ on $S[K]$, we conclude that dN is null on $S[K]$, which shows that $S[K]$ is a Killing horizon, that is, $S[K] = H[K]$. $\qquad\square$

We also establish the following proposition in preparation for the weak rigidity theorem:

Proposition 6.11 *Consider two commuting Killing fields k and m with twist ω_k and ω_m, respectively. Let $X = (m|m)$, $V = -(k|k)$ and $W = (k|m)$, and let ξ and Ω be defined as*

$$\xi = k + \Omega m, \qquad \Omega = -\frac{W}{X}.\qquad (6.9)$$

Then the twist of the vector field (1–form) ξ fulfils the identity

$$\omega_\xi = [(m|\omega_k) - \Omega(k|\omega_m)]\,\frac{m}{X}.\qquad (6.10)$$

Proof By virtue of definition (6.9) we have

$$\xi \wedge d\xi = k \wedge dk + \Omega^2(m \wedge dm) + \Omega(k \wedge dm + m \wedge dk) - k \wedge m \wedge d\Omega.$$

Since $L_k m = [k, m] = 0$, we obtain $dW = di_m k = -i_m dk$ and $dX = -i_m dm$, and thus

$$d\Omega = -\frac{1}{X}\left(dW - \frac{W}{X}dX\right) = \frac{1}{X}i_m\left(dk - \frac{W}{X}dm\right).\qquad (6.11)$$

Using this expression in the last term of the formula for $\xi \wedge d\xi$ now yields

$$\begin{aligned}
\xi \wedge d\xi &= \frac{1}{X}\left[X(k \wedge dk) - W(m \wedge dk) - k \wedge m \wedge i_m dk\right] \\
&+ \frac{W}{X^2}\left[W(m \wedge dm) - X(k \wedge dm) + k \wedge m \wedge i_m dm\right] \\
&= \frac{1}{X}i_m\left[(m \wedge k \wedge dk) + \frac{W}{X}(k \wedge m \wedge dm)\right].
\end{aligned}$$

Using $m \wedge k \wedge dk = 2(m|\omega_k) * 1$, $k \wedge m \wedge dm = 2(k|\omega_m) * 1$ and $i_m * 1 = *m$ then establishes the identity (6.10). $\qquad\square$

Corollary 6.12 *Consider a circular spacetime with Killing fields*
k and m. Then the vector field ξ,

$$\xi = k - \frac{(k|m)}{(m|m)} m ,$$

is hypersurface orthogonal, i.e., $\xi \wedge d\xi = 0$.

Let us now prove the weak rigidity theorem:

Theorem 6.13 *(weak rigidity) Consider a circular spacetime. Let*
$\xi = k + \Omega m$ *with* $\Omega = -W/X$ *and let* $S[\xi]$ *denote the surface*
$\{(\xi|\xi) = 0\}$. *Then* Ω *is constant on* $S[\xi]$, ξ *is a Killing field on*
$S[\xi]$, *and* $S[\xi]$ *is a stationary null surface,* $S[\xi] = H[\xi]$.

Proof We first show that the integrability conditions for k and
m yield $\Omega = $ constant on $S[\xi]$: Using $L_k m = [k, m] = 0$, we obtain
$dW = di_k m = -i_k dm = *(k \wedge *dm)$ and $dX = *(m \wedge dm)$. Hence

$$*d\Omega = -\frac{1}{X} (\xi \wedge *dm) .$$

By definition of ξ and Ω we have $(\xi|m) = 0$ and $(\xi|k) = (\xi|\xi)$. The
above expression therefore yields

$$i_k i_m * d\Omega = \frac{2}{X} [(\xi|\xi) \omega_m - (k|\omega_m) \xi] = 0 \quad \text{on } S[\xi], \quad (6.12)$$

since $(\xi|\xi) = 0$ on $S[\xi]$ and $(k|\omega_m) = 0$. This shows that $d\Omega$ is
a linear combination of k and m on $S[\xi]$, $d\Omega = ak + bm$. For an
arbitrary vector field v orthogonal to the Killing fields we now
have $L_v \Omega = i_v d\Omega = a(v|k) + b(v|m) = 0$ on $S[\xi]$. In addition,
the definition of Ω also implies that $L_k \Omega = L_m \Omega = 0$; hence Ω is
constant on $S[\xi]$.

Since k and m are required to fulfil the integrability conditions
we can apply the previous corollary to conclude that the twist of
ξ vanishes. Since Ω is constant on $S[\xi]$, ξ is a *Killing field on $S[\xi]$*.
Theorem 6.10 then implies that $S[\xi]$ is a null surface.   □

6.4 Properties of Killing horizons

In this section we introduce the surface gravity, κ, and discuss
some important properties of Killing horizons. As before, we use
the quantities K, N and ω to denote the horizon Killing field and

its norm and twist, as defined in eq. (2.9). In theorem 6.10 we argued that the twist of K vanishes on the horizon,

$$N = 0, \quad \omega = 0 \quad \text{on } H[K]. \tag{6.13}$$

The fact that both K and dN are null on $H[K]$ implies that they are proportional. This suggests the following definition:

Definition 6.14 *Let $H[K]$ be the Killing horizon generated by the Killing field K with norm $N = (K|K)$. Then the surface gravity κ of $H[K]$ is defined by the relation*

$$dN = -2\kappa K \quad \text{on } H[K]. \tag{6.14}$$

We first note that the above definition implies that κ is constant along each null geodesic generator of the horizon,

$$L_K \kappa = 0. \tag{6.15}$$

An explicit expression for κ is obtained as follows: Using the definitions of N and ω, we have the identity

$$(dK|dK) * K = i_K(dK \wedge *dK) = -dN \wedge *dK + 2\, dK \wedge \omega,$$

which we evaluate on the horizon: Since $\omega = 0$ on $H[K]$, we obtain after repeated application of eq. (6.14),

$$(dK|dK)*K = -2\kappa *i_K dK = 2\kappa *dN = -4\kappa^2 *K \quad \text{on } H[K],$$

which gives the following expression for the surface gravity in terms of K:

$$\kappa^2 = -\left[\frac{1}{4}(dK|dK)\right]_{H[K]}. \tag{6.16}$$

This yields the following interpretation of κ: Together with the general identity (2.11) and the fact that $N^{-1}(\omega|\omega)$ vanishes on $H[K]$, eq. (6.16) implies that $\kappa^2 = -(4N)^{-1}(dN|dN)\,|_{H[K]}$. Now using $L_K N = 0$ and the Killing equation, we also have $dN = -2\nabla_K K$ (where $(\nabla_K K)^\mu \equiv K^\nu \nabla_\nu K^\mu$). Thus

$$\kappa^2 = -\left[\frac{1}{N}(\nabla_K K \,|\, \nabla_K K)\right]_{H[K]}. \tag{6.17}$$

On the other hand, considering the 4–velocity $u = K/\sqrt{-N}$ and again using $L_K N = 0$, the acceleration $a = \nabla_u u$ of an orbit of K becomes $a = N^{-1}\nabla_K K$. Hence, for a static black hole, $\kappa = \lim(\sqrt{-N}|a|)$ is the limiting value of the force applied at infinity to keep a unit mass at $H[K]$ in place (see, e.g., Wald 1984). Our next aim is to prove the following fundamental proposition:

Proposition 6.15 *Let K be the null generator of a Killing horizon $H[K]$. Then*

$$R(K, K) = 0 \quad on \ H[K]. \tag{6.18}$$

Proof We first note that the derivative operator $\mathcal{D}_{\mu\nu} = K_\mu \nabla_\nu - K_\nu \nabla_\mu$ is tangent to the horizon. It may therefore be applied to equations which are restricted to $H[K]$. For an arbitrary vector field (1–form) α and Killing field K, the contraction $\mathcal{D}_{\mu\nu} \alpha^\nu$ can be written as

$$(K_\mu \nabla_\nu - K_\nu \nabla_\mu) \alpha^\nu = -([Kd^\dagger + di_K] \alpha + i_\alpha \, dK)_\mu. \tag{6.19}$$

We now apply this operator to the definition (6.14) for the surface gravity. Since the last term is algebraic in α, it gives the same contribution when applied on the l.h.s. and the r.h.s. of eq. (6.14). We thus obtain

$$-(Kd^\dagger + di_K) \, dN = 2 \, (Kd^\dagger + di_K) \, (\kappa K)$$
$$\implies \quad K \, \Delta N = -2 \, (K|d\kappa) \, K + 2 \, d(\kappa N)$$
$$\implies \quad K \, \Delta N = -2 \, K \, L_K \kappa + 2 \, N \, d\kappa + 2\kappa \, dN \quad on \ H[K],$$

where we have used $di_K dN = dL_K N = 0$ and $d^\dagger K = 0$. Since the Lie derivative of κ with respect to K vanishes and since $N = 0$ and $dN = -2\kappa K$ on the horizon, we obtain the second explicit expression for the surface gravity,

$$\kappa^2 = - \left[\frac{1}{4} \Delta N \right]_{H[K]}. \tag{6.20}$$

In order to complete the proof, we observe that the general identities (2.18) and (2.11) yield

$$2 \, R(K, K) = (dK|dK) - \Delta N.$$

Taking advantage of the two expressions (6.16) and (6.20) for the surface gravity gives the desired result. $\qquad\square$

It is worth pointing out that the fundamental property (6.18) is also obtained from more general arguments: In order to prove the strong rigidity theorem, one has to show that the vector tangent to the generator of the future event horizon has vanishing shear and expansion. The above result is then obtained from the

Penrose–Raychaudhuri equation for this null tangent vector, without needing to establish the Killing property of the horizon.

Let us eventually note that the surface gravity, κ, measures the extent to which the parametrization of the geodesic congruence $C(s)$ generated by $K = dC/ds$ is not affine. In order to see this, we use the Killing equation to write $(i_K dK)^\mu = 2K^\nu \nabla_\nu K^\mu$, that is, $i_K dK = 2\nabla_K K$. Since $dN = di_K K = -i_K dK$, the definition (6.14) describes nonaffinely parametrized geodesics (see, e.g., Choquet–Bruhat *et al.* 1982)

$$\nabla_K K = \kappa K. \qquad (6.21)$$

Under a reparametrization, $s \mapsto \lambda(s)$, $C(s) = \hat{C}(\lambda) = \hat{C} \circ \lambda(s)$, $\hat{K} = d\hat{C}/d\lambda$, one has ($\lambda' \equiv d\lambda/ds$)

$$\nabla_{\hat{K}} \hat{K} = [\kappa\lambda' - \lambda''](\lambda')^{-2} \hat{K}.$$

Since κ is constant, we see that $\lambda(s) = e^{\kappa s}$ is the *affine* parameter for the geodesics $\hat{C}(\lambda)$ with tangent field $\hat{K}$:

$$\nabla_{\hat{K}} \hat{K} = 0, \quad \hat{K} = (\kappa\lambda)^{-1} K. \qquad (6.22)$$

This enables one also to conclude that K^μ vanishes along each geodesically complete null–generator of $H[K]$, provided that $\kappa \neq 0$. Thus, $K^\mu = 0$ on a spacelike cross section $\mathcal{F}$, implying that the horizon is of bifurcate type (Kay and Wald 1991):

Corollary 6.16 *A regular (i.e., $\kappa \neq 0$) Killing horizon with geodesically complete null–generators is a bifurcate Killing horizon.*

6.5 The topology of the horizon

This section is devoted to an outline of the proof of the topology theorem as given in Hawking and Ellis (1973). The theorem is of fundamental interest for the black hole uniqueness results since the latter only apply to configurations for which the connected components of the event horizon have spherical topology.

The basic idea is to establish a connection between the Euler characteristic, χ, of the 2–surface $\mathcal{H}$ and the expansion of the outgoing null geodesic congruence. Assuming that the stress–energy tensor obeys the dominant energy condition, it is then argued that for $\chi(\mathcal{H}) < 0$ it is possible to deform $\mathcal{H}$ outwards into $J^-(\mathcal{J}^+)$ such that the outgoing null geodesics have nonpositive expansion.

In order to avoid a contradiction to the fact that no outer trapped surface can intersect $J^-(\mathcal{J}^+)$ (see the next chapter), one concludes that the Euler characteristic is non–negative, that is $\chi(\mathcal{H}) = 0$ or $\chi(\mathcal{H}) = 2$. Hence, $\mathcal{H}$ has either toroidal or spherical topology.

The exclusion of the toroidal case requires more involved arguments. In fact, the problem was settled only recently by different means. Here we present the original reasoning given by Hawking, which excludes the cases with strictly negative Euler characteristic. The reader is referred to the end of this section for some comments on recent progress toward the final proof of the theorem.

Theorem 6.17 *Consider a stationary, regular predictable space-time. Then, provided that the stress–energy tensor satisfies the dominant energy condition, each connected component $\mathcal{H}$ of the intersection of the horizon H with a Cauchy surface Σ is homeomorphic to a 2–sphere.*

Proof Let us denote the null–generator of the horizon by K and the second future directed null vector orthogonal to the horizon by n; normalized such that $(K|n) = -1$. The induced metric on the 2–surface $\mathcal{H}$ is

$$\beta = g + K \otimes n + n \otimes K. \tag{6.23}$$

According to the Gauss–Bonnet theorem (see, e.g., Kobayashi and Nomizu 1969, Vol. 2) the Euler characteristic of a compact orientable 2–dimensional Riemannian manifold $\mathcal{H}$ is given by the integral

$$\chi(\mathcal{H}) = \frac{1}{2\pi} \int_{\mathcal{H}} K_G \tilde{\eta}, \tag{6.24}$$

where K_G and $\tilde{\eta}$ denote the Gauss curvature and the volume form of $\mathcal{H}$, respectively. Recall that $\chi(\mathcal{H}) \leq 0$ for all compact orientable 2–surfaces which are not homeomorphic to S^2, and $\chi(S^2) = 2$. The equations of Gauss and Codazzi imply that the Gauss curvature of the 2–surface $\mathcal{H}$ can be expressed in terms of the Riemann tensor as (see, e.g., Spivak 1979)

$$K_G = \frac{1}{2} R_{\mu\nu\sigma\rho} \beta^{\mu\sigma} \beta^{\nu\rho}, \tag{6.25}$$

provided that $\mathcal{H}$ has vanishing shear and expansion. Since this is the case for the horizon, one can use the above expressions to write the Gauss curvature in terms of K and n. Taking advantage of the symmetry properties of the Riemann tensor, one finds

$$K_G = R + 4R(K,n) - 2R(K,n,K,n), \qquad (6.26)$$

where $R(K,n,K,n) = R_{\mu\nu\sigma\rho}K^\mu n^\nu K^\sigma n^\rho$ and R is the Ricci scalar. Together with the normalization $(K|n) = -1$, the (K,n)–component of Einstein's equations becomes $16\pi T(K,n) = 2R(K,n)+R$. This yields the following formula for the Euler characteristic (6.24)

$$\chi(\mathcal{H}) = 4\int_\mathcal{H} T(K,n)\,\tilde\eta + \frac{1}{2\pi}\int_\mathcal{H}[R(K,n) - R(K,n,K,n)]\,\tilde\eta\,.$$
$$(6.27)$$

In order to evaluate the second integral, we consider the null geodesic congruence tangent to n, parametrized by s, say. This defines a family $\mathcal{H}(s)$ of 2–surfaces, with $\mathcal{H}$ being the surface $s = 0$. The variation of the expansion rate $\theta = \beta^\nu_{\ \mu}\nabla_\nu K^\mu$ with respect to s is

$$\frac{d\theta}{ds} = n^\sigma\nabla_\sigma\left(\beta^\nu_{\ \mu}\nabla_\nu K^\mu\right). \qquad (6.28)$$

Introducing the field $p^\mu = -n^\sigma\beta^{\mu\nu}\nabla_\nu K_\sigma$, a lengthy but straightforward derivation yields the following expression for the second integrand in eq. (6.27) (use the general relation (1.4) in order to commute second covariant derivatives):

$$R(K,n) - R(K,n,K,n)$$
$$= -\frac{d\theta}{ds} - \beta^\mu_\sigma\beta^\nu_\rho\nabla_\mu n^\rho\nabla_\nu K^\sigma - \tilde d^\dagger p + (p|p)\,, \qquad (6.29)$$

where $\tilde d^\dagger p = -\beta^{\mu\nu}\nabla_\nu\beta_\mu$. In order to proceed, one observes that both the normalization condition $(K|n) = -1$ and the induced metric β are invariant with respect to a scaling transformation of the form

$$K \to K' = e^f K\,, \qquad n \to n' = e^{-f} n\,. \qquad (6.30)$$

Making again use of the fact that the horizon has vanishing shear and expansion, we see that the second term on the r.h.s. of eq. (6.29) vanishes on $\mathcal{H}$. Hence

$$R(K,n) - R(K,n,K,n) + \tilde d^\dagger p' + \tilde d^\dagger df$$
$$= (p'|p') - \frac{d\theta'}{ds'}\,, \qquad \text{on } \mathcal{H}\,, \qquad (6.31)$$

where $p'^\mu \equiv p^\mu + \beta^{\mu\nu} \nabla_\nu f$, θ' is the expansion rate with respect to K' and s' parametrizes the null congruence tangent to n'. Integrating this equation over $\mathcal{H}$ and using the fact that the third and fourth integrand on the l.h.s. are exact 2–forms ($\tilde{d}^\dagger p' \, \tilde{\eta} = \tilde{*} \tilde{d}^\dagger p' = d\tilde{*}p'$), eq. (6.27) finally becomes

$$\chi(\mathcal{H}) = 4 \int_{\mathcal{H}} T(K,n) \, \tilde{\eta} + \frac{1}{2\pi} \int_{\mathcal{H}} \left[(p'|p') - \frac{d\theta'}{ds'} \right] \tilde{\eta}. \qquad (6.32)$$

According to a theorem of Hodge (1959), there exists a function f such that the l.h.s. of eq. (6.31) assumes a constant value. Hence, denoting the area of $\mathcal{H}$ by $\mathcal{A}$, the Euler characteristic finally becomes

$$\chi(\mathcal{H}) = 4 \int_{\mathcal{H}} T(K,n) \, \tilde{\eta} + \frac{\mathcal{A}}{2\pi} \left[(p'|p') - \frac{d\theta'}{ds'} \right]_{s=0}, \qquad (6.33)$$

since the term in brackets is constant on the horizon.

Since K and n are future directed and null, the dominant energy condition implies that $T(K,n)$ is not negative. Hence, a negative Euler number can be obtained only if

$$\frac{d\theta'}{ds'} > (p'|p') \geq 0, \quad \text{for } s = 0, \qquad (6.34)$$

that is, if $d\theta'/ds'$ is positive over the entire horizon. Suppose that f can be chosen such that this is the case. Then, for small negative values of s, one obtains a 2–surface in $J^-(\mathcal{J}^+)$ with converging outgoing null geodesics, contradicting the fact that no outer trapped surfaces can intersect $J^-(\mathcal{J}^+)$. This proves that the Euler number is either positive (i.e., $\chi(\mathcal{H}) = 2$) or zero. $\qquad \square$

It remains to exclude the toroidal case, $\chi(\mathcal{H}) = 0$. As already mentioned, this was achieved only recently by Chruściel and Wald (1994b). Their proof takes advantage of the topological censorship theorem (Friedman *et al.* 1993), in order to establish that the domain of outer communications is simply connected if the null energy condition holds. (This excludes, of course, *all* non-spherical topologies.) In addition, the reasoning does not require the same amount of differentiability (on the horizon) as the traditional proof.

A similar result was recently obtained by Jacobson and Venkataramani (1995). Also making use of the topological censorship theorem, they were able to establish that the horizon cross section consists of a disjoint union of 2–spheres. Their arguments, which do not require stationarity, are based on the averaged null energy condition and global hyperbolicity of the domain of outer communications. The reader who is interested in the modern approach to the topology theorem should also consult Masood–ul–Alam (1987), Galloway (1993, 1994) and Lindblom and Masood–ul–Alam (1994).

7

The four laws of black hole physics

The area theorem is probably one of the most important results in classical black hole physics. It asserts that (under certain conditions which we specify below) the area of the event horizon of a predictable black hole spacetime cannot decrease. This result bears a resemblance to the second law of thermodynamics. The analogy is reinforced by the similarity of the mass variation formula to the first law of ordinary thermodynamics. Within the *classical* framework the analogy is basically of a formal, mathematical nature. There exists, for instance, no physical relationship between the surface gravity, κ, and the classical temperature of a black hole, which must be assigned the value of absolute zero. Nevertheless, on account of the Hawking effect, the relationship between the laws of black hole physics and thermodynamics gains a deep *physical* significance: The temperature of the black–body spectrum of particles created by a black hole is $\kappa/2\pi$. This also sheds light on the analogy between the entropy and the area of a black hole.

The Killing property of a stationary event horizon implies that its surface gravity is constant. If the Killing fields are integrable (that is, in static or circular spacetimes), the zeroth law of black hole physics is a purely geometrical property of Killing horizons. Otherwise, it is a consequence of Einstein's equations and the dominant energy condition.

The Komar expression for the mass of a stationary spacetime provides a formula giving the mass in terms of the total angular momentum, the angular velocity, the surface gravity and the area of the horizon. Variation of this expression between neighboring stationary states yields the so–called first law of black hole physics. It was originally derived by Bardeen *et al.* (1973) for the vacuum case and was subsequently generalized by Carter (1973a, 1979) to electrovac black hole spacetimes. More recently, the first law has also been established for new black hole solutions with non–Abelian gauge or Skyrme fields (Sudarsky and Wald 1992, 1993, Heusler and Straumann 1993a, 1993b). Moreover, Iyer, Sudarsky and Wald were able to show that the first law is valid for a considerably larger class of perturbations than the ones admitted in the original derivation (Sudarsky and Wald 1992, Wald 1993a, Iyer and Wald 1994). In addition, Wald (1993b) gave a derivation of the first law which is based on symmetry considerations and applies to arbitrary theories of gravity which arise from diffeomorphism–invariant Lagrangians.

This chapter is organized as follows: The first section is devoted to the *zeroth law*. The traditional reasoning establishes the uniformity of κ for arbitrary Killing horizons if the matter is subject to the dominant energy condition. We also present two newer arguments for $\kappa =$ constant. They are, however, restricted to the cases where either the horizon is of *bifurcate* type (Kay and Wald 1991) or spacetime is static or circular.

The second section deals with the first law of black hole physics. Following the traditional approach of Bardeen *et al.* (1973), the mass variation formula for two neighboring stationary black hole configurations is derived from the generalized Smarr formula. In view of later applications, the matter model is not specified. We shall find that - besides the usual vacuum terms - the total mass and its variation can be expressed in terms of the matter Lagrangian, provided that the latter does not involve derivatives of the metric (Heusler and Straumann 1993a). As an application, we give the mass variation formula for electrovac spacetimes.

A brief discussion of the area increase theorem is given in the third section. We conclude this chapter with some comments on the third law and on the relationship between the laws of black hole physics and the laws of ordinary thermodynamics.

7.1 The zeroth law

It is an interesting fact that the surface gravity plays a similar role in the classical theory of black holes as the temperature does in ordinary thermodynamics. Since the latter is constant for a body in thermal equilibrium, the result that κ is uniform over the horizon is usually called the zeroth law of black hole mechanics. As already mentioned, this formal analogy gains a deeper significance by the fact that κ is (up to a factor of $\hbar/2\pi$) equal to the temperature of the black–body spectrum of the Hawking radiation.

In this section we derive the well–known result that κ assumes a constant value on a Killing horizon (Bardeen *et al.* 1973). We first present the traditional approach, which is valid for arbitrary Killing horizons, provided that Einstein's equations hold with matter satisfying the dominant energy condition (see, e.g., Wald 1984). As was pointed out by Kay and Wald (1991), there exists a much simpler proof which does not involve Einstein's equations if the Killing horizon is of bifurcate type. In fact, this is probably not a strong restriction since, as was recently pointed out by Rácz and Wald (1992, 1995) all "physically relevant" Killing horizons are either bifurcate horizons or degenerate (i.e., have vanishing surface gravity). There exists yet another method of establishing the zeroth law: The proof is of purely geometric nature and considerably simpler than the traditional argument. It has, however, the drawback of being restricted to static or circular configurations, since it is based on integrability conditions.

Let us now derive the first law under the assumption that Einstein's equations, with matter satisfying the dominant energy condition, are satisfied. To start, we note that the former, together with the fundamental identity (6.18) implies that the component $T(K, K)$ of the stress–energy tensor vanishes on the horizon. Thus, the 1–form $T(K)$ is perpendicular to K and therefore spacelike or null on $H[K]$. On the other hand, the dominant energy condition requires that $T(K)$ is timelike or null. Hence, $T(K)$ is null on the horizon and therefore also proportional to K. Using Einstein's equations again yields

$$R(K) \wedge K = 0 \quad \text{on } H[K]. \tag{7.1}$$

The main task in establishing the zeroth law is to derive the identity

$$K \wedge d\kappa = -K \wedge R(K) \quad \text{on } H[K]. \tag{7.2}$$

In combination with the consequence (7.1) of the field equations, this formula proves that κ is constant, since the l.h.s. is equal to the tangential derivative $\mathcal{D}_{\mu\nu}$ of κ,

$$\mathcal{D}_{\mu\nu}\kappa = (K_\mu\nabla_\nu - K_\nu\nabla_\mu)\kappa = (K \wedge d\kappa)_{\mu\nu} \quad \text{on } H[K]. \quad (7.3)$$

In order to prove the identity (7.2), which is valid for arbitrary Killing horizons, one first notes that the Killing horizon property $K \wedge dK = 0$ (6.13) and the definition (6.14) can be written in the form

$$\mathcal{D}_{\mu\nu}K_\sigma = K_\sigma\nabla_\nu K_\mu, \quad \kappa K_\sigma = K^\rho\nabla_\rho K_\sigma \quad \text{on } H[K], \quad (7.4)$$

respectively. As an immediate consequence of these expressions one has

$$\kappa\,\mathcal{D}_{\mu\nu}K_\sigma = \kappa\,K_\sigma\nabla_\nu K_\mu = K^\rho\nabla_\nu K_\mu\nabla_\rho K_\sigma = \nabla_\rho K_\sigma\mathcal{D}_{\mu\nu}K^\rho,$$

from which one finds that two terms in the tangential derivative of eq. (6.14) cancel:

$$K_\sigma\,\mathcal{D}_{\mu\nu}\kappa + \kappa\,\mathcal{D}_{\mu\nu}K_\sigma = \nabla_\rho K_\sigma\mathcal{D}_{\mu\nu}K^\rho + K^\rho\mathcal{D}_{\mu\nu}\nabla_\rho K_\sigma.$$

By virtue of the explicit expression for $\mathcal{D}_{\mu\nu}$ and the general consequence $\nabla_\nu\nabla_\rho K_\sigma = -R_{\rho\sigma\nu}{}^\beta K_\beta$ of eq. (1.4) for Killing fields, we obtain the following equation for the tangential derivative of κ:

$$K_\sigma\,\mathcal{D}_{\mu\nu}\kappa = 2\,K^\rho K_{[\mu}\nabla_{\nu]}\nabla_\rho K_\sigma = 2\,K_{[\mu}R_{\nu]\beta\sigma\rho}K^\beta K^\rho. \quad (7.5)$$

In order to complete the derivation, one has to verify the identity (valid on $H[K]$)

$$K_{[\mu}R_{\nu]\beta\sigma\rho}K^\beta K^\rho = -K_{[\mu}R_{\nu]}^{\ \rho}K_\rho K_\sigma, \quad (7.6)$$

which is obtained by applying the tangential operator $\mathcal{D}_{\alpha\beta}$ to the first equation (7.4) and contracting over σ and β (see Wald 1984). Substituting this expression in eq. (7.5) eventually gives the desired result (7.2).

This completes the derivation of the zeroth law of black hole dynamics in its original form (Bardeen *et al.* 1973):

Theorem 7.1 *(zeroth law) Let $H[K]$ be a Killing horizon. Then the surface gravity is constant on $H[K]$, provided that Einstein's equations hold with matter satisfying the dominant energy condition.*

As already mentioned, a stronger form of the zeroth law is obtained if the generators of the horizon are assumed to be geodesically complete (Kay and Wald 1991, Wald 1992). In this case, Einstein's equations and the dominant energy condition are not required. As mentioned at the end of section 6.4, geodesic completeness implies the existence of a bifurcation 2–surface $\mathcal{F}$ in a neighborhood of a generator along which κ does not vanish. (Recall that κ is constant *along* the generators.) In order to show that κ cannot vary between generators either, one considers the derivative of the explicit expression (6.16) for κ in a direction tangent to $\mathcal{F}$, Z^μ, say. Using again the general property $\nabla_\mu \nabla_\nu K_\sigma = -R_{\nu\sigma\mu}{}^\rho K_\rho$, this yields

$$\kappa L_Z \kappa = -\frac{1}{2} Z^\mu \left(\nabla_\mu \nabla_\nu K_\sigma \right) \left(\nabla^\nu K^\sigma \right)$$

$$= \frac{1}{2} K^\rho Z^\mu R_{\nu\sigma\mu\rho} \nabla^\nu K^\sigma = 0, \qquad (7.7)$$

since K^ρ vanishes on the bifurcation 2–surface $\mathcal{F}$. Hence, $L_K \kappa = 0$ *and* $L_Z \kappa = 0$, implying the desired result (Kay and Wald 1991):

Theorem 7.2 *(zeroth law; for bifurcate horizons) Let $H[K]$ be a bifurcate Killing horizon. Then the surface gravity is constant on $H[K]$.*

In this connection we also refer to the work of Rácz and Wald (1992), which shows that for any connected Killing horizon with nonconstant surface gravity, some incomplete null–generators terminate in a curvature singularity. In addition they prove that - if κ is assumed to be constant and nonvanishing over a Killing horizon $H[K]$ - the spacetime can be extended such that $H[K]$ is a subset of a bifurcate Killing horizon. As the authors emphasize, these results suggest that all physically relevant Killing horizons are either of bifurcate type or degenerate.

The following observation implies an alternative formulation of the zeroth law: The traditional proof of the zeroth law is based on the identity (7.2) and the observation that Einstein's equations for matter satisfying the dominant energy condition imply $R(K) \wedge K = 0$ on $H[K]$. However, this is equivalent to the condition $d\omega_K = 0$ on $H[K]$. We thus have the following theorem (see also Rácz and Wald 1995):

Theorem 7.3 *(zeroth law; for closed twist) Let $H[K]$ be a Killing horizon generated by K. Let $d\omega_K = 0$ on $H[K]$. Then the surface gravity is constant on $H[K]$.*

One way to achieve $d\omega = 0$ on the horizon is, of course, to require that ω vanishes identically (i.e., not only *on* the horizon). Thus, in a static spacetime, the surface gravity is automatically constant. In fact, the same is true in the circular case. Hence, we obtain the fourth form of the zeroth law:

Theorem 7.4 *(zeroth law; for static or circular spacetimes) Let $H[K]$ be an arbitrary Killing horizon and let the domain of outer communications be static or circular. Then the surface gravity is constant on $H[K]$.*

As already mentioned, this theorem is, at least in the static case, an immediate consequence of the identity (7.2). However, since the derivation of this identity is not especially illuminating, we shall now present a direct proof of the above theorem. The reasoning basically consists in the evaluation of the following identity on the horizon:

Proposition 7.5 *Let K be a Killing field (with twist ω and norm N) and let Z be an arbitrary vector field (1–form). Then*

$$2di_Z i_K * \omega = -i_Z(dN \wedge dK) + d(Z|K) \wedge dN + K \wedge di_Z dN. \quad (7.8)$$

Proof Using the definition (2.9) for ω and the fact that $i_K dK = -di_K K = -dN$, we obtain the identity $2i_K * \omega = NdK + K \wedge dN$. The inner product of this with respect to Z becomes

$$2i_Z i_K * \omega = Ni_Z dK + (Z|K)dN - K \wedge i_Z dN\,.$$

Applying the exterior derivative on this formula and writing $dN \wedge i_Z dK - dK \wedge i_Z dN$ as $-i_Z(dN \wedge dK)$ then establishes the desired result. $\qquad\qquad\square$

Using the above proposition, it is now not difficult to give the following direct proof of theorem 7.4:

Proof (of theorem 7.4) Let us start with the static case and evaluate the identity (7.8) on the horizon. Since the null–generator Killing field K coincides with the stationary Killing field k, ω vanishes *identically*. Hence, the derivative of $i_Z i_K * \omega$ is zero as well, implying that the l.h.s. of eq. (7.8) vanishes. Using $dN = -2\kappa K$, the first term on the r.h.s. vanishes as well since it becomes proportional to $K \wedge dK$ (which is always zero on a Killing horizon). Again taking advantage of the definition of κ, we thus obtain from eq. (7.8)

$$
\begin{aligned}
0 &= d(Z|K) \wedge (\kappa K) + K \wedge di_Z (\kappa K) \\
&= (Z|K) K \wedge d\kappa \quad \text{on } H ,
\end{aligned}
\tag{7.9}
$$

which establishes $K \wedge d\kappa = 0$ on the horizon. (For instance, let $Z = n$ be the second null vector orthogonal to the horizon, normalized such that $(n|K) = -1$.) As we have argued above, $K \wedge d\kappa = 0$ is sufficient to conclude that κ is constant.

The proof of the circular case ($K \equiv \ell$) requires an additional step, since $\omega \equiv \omega_\ell$ does not vanish *identically*. However, as a consequence of the integrability conditions, the quantity $m \wedge *\omega = (m|\omega)\eta$ does vanish identically, since

$$
(m|\omega) = (m|\omega_k) - \Omega_H (k|\omega_m) = 0 .
$$

We therefore consider the vanishing quantity $di_Z i_\ell(m \wedge *\omega)$, which we write in the form

$$
\begin{aligned}
di_Z i_\ell(m \wedge *\omega) &= d\left[(l|m)i_Z * \omega\right] - d\left[(Z|m)i_\ell * \omega\right] \\
&\quad - dm \wedge i_Z i_\ell * \omega + m \wedge di_Z i_\ell * \omega .
\end{aligned}
$$

Let us evaluate the r.h.s. on the horizon. The first term vanishes, since *both* $(\ell|m)$ and $*\omega = \frac{1}{2}\ell \wedge d\ell$ vanish on the Killing horizon generated by ℓ. The only contribution which does not manifestly vanish in the second term comes from applying the derivative operator on $i_\ell * \omega$. However, $di_\ell * \omega = \frac{1}{2}d[Nd\ell + \ell \wedge dN] = dN \wedge d\ell$, which also vanishes on the horizon, since dN is proportional to ℓ and $\ell \wedge d\ell = 0$ on $H[\ell]$. The third term is manifestly zero since $\omega = 0$ on $H[\ell]$. Since the l.h.s. vanishes *identically*, we obtain for the last term

$$
m \wedge di_Z i_\ell * \omega = 0 \quad \text{on } H[\ell] .
$$

Now we can apply the proposition and proceed exactly in the same way as in the static case. This yields the conclusion $m \wedge \ell \wedge d\kappa = 0$, or equivalently

$$
m \wedge k \wedge d\kappa = 0 \quad \text{on } H[\ell] .
$$

This is, however, sufficient to conclude that κ is constant on the horizon. $\qquad\square$

7.2 The first law

As we have seen in section 2.4, the total gravitating mass of a stationary asymptotically flat spacetime can be expressed in terms of the Killing field k by the Komar integral,

$$M = -\frac{1}{8\pi} \int_{S^2_\infty} *dk .$$ (7.10)

Applying Stokes' theorem and using $k = \ell - \Omega_H m$ (see eq. (6.1)), the above expression yields a formula for the mass of a stationary (and axisymmetric, if $\Omega_H \neq 0$) black hole spacetime in terms of the area $\mathcal{A}$, the surface gravity κ, the angular velocity Ω_H and the angular momentum J_H of the event horizon. This generalizes an expression first derived by Smarr (1973) for the Kerr metric. On the basis of this formula, Bardeen *et al.* (1973) computed the infinitesimal difference between two neighboring stationary black hole solutions (see also Carter 1979). In the vacuum case, this yields

$$\delta M = \frac{1}{8\pi} \kappa \, \delta\mathcal{A} + \Omega_H \, \delta J_H ,$$ (7.11)

which stands in an obvious analogy to the first law of thermodynamics, $\delta E = T\delta S + p\delta V$. We have already seen that the zeroth law of black hole physics indicates a relationship between the surface gravity κ and the temperature T. Hence, the variation formula (7.11) suggests an analogy between the area $\mathcal{A}$ of the event horizon and the entropy S. In fact, the analogy is reinforced by the second law of black hole physics (see below), asserting that the area cannot decrease in a classical process.

In this section we present an outline of the traditional approach to the first law. By now, there also exists a more general derivation of the mass variation formula (Sudarsky and Wald 1992, Wald 1993a, Iyer and Wald 1994). The reasoning is based on the Hamiltonian formalism of general relativity (Arnowitt *et al.* 1962; see also Fischer and Marsden 1972, 1976, 1979) and does not require stationary perturbations. Moreover, the arguments can be generalized to arbitrary diffeomorphism–invariant theories of gravity which are obtained from a Lagrangian (Iyer and Wald 1994). The reader is also referred to Wald (1994) for an introduction into these new methods.

We start by deriving the mass formula. Using Stokes' theorem and the Ricci identity (2.7) for the Killing field k, the Komar expression (7.10) becomes

$$M = M_H - \frac{1}{4\pi} \int_\Sigma *R(k).$$ (7.12)

As usual, Σ denotes a spacelike hypersurface extending from the inner boundary $\mathcal{H}$ (the intersection of Σ with the horizon H) to spacelike infinity S^2_∞. In order to evaluate the expression for M_H we use $k = \ell - \Omega_H m$ ($\Omega_H = 0$ in the nonrotating case):

$$M_H = -\frac{1}{8\pi} \int_\mathcal{H} *dk = -\frac{1}{8\pi} \int_\mathcal{H} *d\ell + 2\Omega_H J_H.$$ (7.13)

Here we have also used the fact that Ω_H is constant and the definition

$$J_H = \frac{1}{16\pi} \int_\mathcal{H} *dm$$ (7.14)

for the angular momentum of the horizon. The surface contribution from the horizon can be evaluated by using the fact that for an arbitrary 2–form Λ we have

$$\int_\mathcal{H} *\Lambda = \int_\mathcal{H} i_{n_\ell} i_\ell \Lambda \, dA,$$ (7.15)

where n_ℓ denotes the second future–directed null vector orthogonal to the horizon, normalized such that $(n_\ell | \ell) = -1$. Hence, the boundary integral in eq. (7.13) becomes

$$\int_\mathcal{H} *d\ell = \int_\mathcal{H} i_{n_\ell} i_\ell d\ell \, dA = -\int_\mathcal{H} (n_\ell | dN) \, dA$$

$$= -2\kappa \int_\mathcal{H} dA = -2\kappa \mathcal{A},$$ (7.16)

where $N = (\ell | \ell)$ and where we have used $L_\ell \ell = 0$, the definition (6.14) of the surface gravity and the fact that κ is constant. Substituting the boundary term in eq. (7.13) by this expression and using the result in eq. (7.12) gives the mass formulae

$$M = \frac{1}{4\pi} \kappa \mathcal{A} + 2\Omega_H J_H - \frac{1}{4\pi} \int_\Sigma *R(k),$$ (7.17)

$$M = \frac{1}{4\pi} \kappa \mathcal{A} + 2\Omega_H J - \frac{1}{4\pi} \int_\Sigma *R(\ell),$$ (7.18)

where we have replaced J_H by $J - \frac{1}{8\pi} \int_\Sigma *R(m)$ (see eq. (2.74)) in order to obtain the second expression.

Following Bardeen *et al.* (1973), we can now compute the mass variation between two neighboring stationary (and axisymmetric) configurations. In contrast to this "equilibrium state version", there also exists a "physical process version" of the derivation. The latter can, for instance, be found in Wald (1994).

To start, we note that the gauge may be chosen such that the Killing fields k^μ and m^μ are the same for both solutions,

$$\delta k^\mu = 0, \qquad \delta k_\mu = h_{\mu\nu} k^\nu, \qquad (7.19)$$

$$\delta m^\mu = 0, \qquad \delta m_\mu = h_{\mu\nu} m^\nu, \qquad (7.20)$$

and thus

$$\delta \ell^\mu = m^\mu \delta \Omega_H, \qquad \delta \ell_\mu = m_\mu \delta \Omega_H + h_{\mu\nu} \ell^\nu. \qquad (7.21)$$

Here we have introduced the variation of the metric, $h_{\mu\nu}$, defined by

$$h_{\mu\nu} = \delta g_{\mu\nu}, \qquad h^\mu{}_\nu = g^{\mu\sigma} h_{\sigma\nu}. \qquad (7.22)$$

Requiring, in addition, that for both solutions the event horizon is located at the same position, we also see that the 1-form $\delta\ell$ satisfies

$$\ell \wedge \delta\ell = 0, \qquad L_\ell \delta\ell = 0. \qquad (7.23)$$

As a consequence of these properties, the following useful relations hold on the horizon $H[\ell]$:

$$n^\mu \delta\ell_\nu + \delta n^\mu \ell_\nu = 0, \qquad (7.24)$$

$$i_X i_Y d(\delta\ell) = 0, \qquad (7.25)$$

where X and Y denote two arbitrary vector fields tangent to the horizon with $(X|\ell) = (Y|\ell) = 0$. As before, $n \equiv n_\ell$ is the second null vector orthogonal to the horizon, normalized such that $(\ell|n) = -1$. Equation (7.24) is an immediate consequence of $\delta\ell = f\ell$, $\delta n = \tilde{f} n$ and the normalization condition, which implies that the functions f and $-\tilde{f}$ are equal. In order to derive eq. (7.25), one first notes that the orthogonality of ℓ to X and Y, together with $\ell \wedge d\ell = 0$ on $H[\ell]$, implies $i_X i_Y d\ell = 0$ (since $0 = i_X i_Y (\ell \wedge d\ell) = \ell i_X i_Y d\ell$). Thus, one obtains on the horizon

$$
\begin{aligned}
i_X i_Y d(\delta\ell) &= i_X i_Y (df \wedge \ell + f \, d\ell) \\
&= (\ell|X) L_Y f - (\ell|Y) L_X f + f \, i_X i_Y d\ell = 0,
\end{aligned}
$$

which establishes eq. (7.25). We are now in a position to prove the following proposition:

Proposition 7.6 *Consider the vector field h^μ, defined in terms of the covariant derivative of the variation of the metric by*

$$h^\mu = \nabla^{[\nu} h^{\mu]}_{\ \nu}. \tag{7.26}$$

Then the variation of the surface gravity between two neighboring stationary (and axisymmetric) black hole solutions is

$$\delta\kappa = -(\ell|h) + \frac{1}{2}\delta\Omega_H \, i_\ell i_n \, dm. \tag{7.27}$$

Proof Let us derive this result in component notation. We first note that

$$n^\mu \nabla_\mu (\ell^\nu \delta\ell_\nu) = [\ell^\nu n^\mu + \ell^\mu n^\nu]\nabla_\mu \delta\ell_\nu + 2n^\mu \, \delta\ell_\nu \nabla_\mu \ell^\nu, \tag{7.28}$$

$$n^\mu \nabla_\mu (\ell_\nu \delta\ell^\nu) = [\ell_\nu n^\mu - \ell^\mu n_\nu] \nabla_\mu \delta\ell^\nu. \tag{7.29}$$

Here we have added the vanishing term $n^\mu \, (L_\ell \delta\ell)_\mu$, i.e.,

$$n^\mu \, (\, \ell^\nu \, \nabla_\nu \delta\ell_\mu + \delta\ell_\nu \, \nabla_\mu \ell^\nu\,) = 0, \tag{7.30}$$

on the r.h.s. of eq. (7.28), and subtracted the same term on the r.h.s. of eq. (7.29). Contracting the definition (6.14) for κ with n gives the expression

$$2\,\kappa = i_n \, dN = n^\mu \, \nabla_\mu (\ell^\nu \, \ell_\nu)$$

for the surface gravity. Together with eqs. (7.28) and (7.29) this yields the variation formula

$$\begin{aligned}
2\,\delta\kappa &= n^\mu \, \nabla_\mu (\ell^\nu \delta\ell_\nu) + n^\mu \, \nabla_\mu (\ell_\nu \delta\ell^\nu) + \delta n^\mu \, \nabla_\mu (\ell^\nu \ell_\nu) \\
&= [\ell^\nu n^\mu + \ell^\mu n^\nu]\nabla_\mu\delta\ell_\nu + 2\,[n^\mu\delta\ell_\nu + \delta n^\mu \ell_\nu]\nabla_\mu \ell^\nu \\
&\quad + [\ell_\nu n^\mu - n_\nu \ell^\mu]\nabla_\mu \delta\ell^\nu, \tag{7.31}
\end{aligned}$$

where eq. (7.24) implies that the second term on the r.h.s. vanishes. In addition, eq. (7.25) shows that the first term on the r.h.s. is equal to minus the covariant divergence of $\delta\ell_\mu$,

$$\nabla_\mu (\delta\ell_\nu) \, [n^\mu l^\nu + n^\nu l^\mu] = -\nabla^\mu (\delta\ell_\mu), \tag{7.32}$$

which can be further evaluated by using eq. (7.21):

$$\nabla^\mu(\delta\ell_\mu) = \nabla^\mu(h^\nu_{\ \mu}\ell_\nu + \delta\Omega_H m_\mu) = \ell_\nu \nabla^\mu h^\nu_{\ \mu} = 2\,(\ell|h). \tag{7.33}$$

Here we have used the Killing equations for ℓ and m and the fact that Ω_H is constant and $L_\ell \, \mathrm{tr}(h) = 0$. Finally, taking advantage of eq. (7.21) in order to evaluate the last term in eq. (7.31), we obtain

$$\delta\kappa = -(\ell|h) + \delta\Omega_H \, \ell^{[\nu} n^{\mu]} \nabla_\mu m_\nu,$$

which completes the proof. $\qquad\square$

Corollary 7.7 *Let κ, $\mathcal{A}$, J_H and Ω_H be the surface gravity, the surface area, the angular momentum and the angular velocity of the horizon, respectively. Let h denote the 1–form assigned to the vector field h^μ defined in eq. (7.26). Then*

$$- \mathcal{A}\,\delta\kappa = 4\pi\,\delta M + 8\pi\,J_H\,\delta\Omega_H + \int_\Sigma d^\dagger h * k, \qquad (7.34)$$

where Σ denotes a spacelike hypersurface extending from $\mathcal{H}$ to S_∞^2, and M is the total mass.

Proof The corollary is the integral version of the formula (7.27) for the variation of κ. Using the expression (7.15) for the evaluation of a 2–form on the horizon yields

$$\int_\mathcal{H} *dm = - \int_\mathcal{H} (i_\ell i_n \, dm) \, d\mathcal{A}, \qquad (7.35)$$

$$\int_\mathcal{H} *(k \wedge h) = \int_\mathcal{H} i_n i_\ell (k \wedge h) \, d\mathcal{A} = \int_\mathcal{H} (h|\ell) \, d\mathcal{A}, \qquad (7.36)$$

since $(k|\ell) = 0$ and $(k|n) = -1$ on $H[\ell]$. Integrating eq. (7.27) over $\mathcal{H}$ and using the fact that κ and Ω_H are constant, the above expressions yield

$$- \delta\kappa \int_\mathcal{H} d\mathcal{A} = \int_\mathcal{H} *(k \wedge h) + \frac{1}{2}\delta\Omega_H \int_\mathcal{H} *dm, \qquad (7.37)$$

where the last term is equal to $8\pi J_H \delta\Omega_H$. In order to evaluate the first term on the r.h.s. we apply Stokes' theorem and use the Killing field identity (2.4) and $L_k h = 0$ to write $d * (k \wedge h) = *d^\dagger(k \wedge h) = -d^\dagger h * k$. This gives

$$\int_\mathcal{H} *(k \wedge h) = \int_{S_\infty^2} *(k \wedge h) - \int_\Sigma d * (k \wedge h) = 4\pi\,\delta M + \int_\Sigma d^\dagger h * k.$$

Here we have also used the formula

$$\delta M = \frac{1}{4\pi} \int_{S_\infty^2} k^\mu \, \nabla^{[\sigma} h^{\nu]}_{\ \ \sigma} \, dS_{\mu\nu}, \qquad (7.38)$$

which is obtained from the asymptotic expansion of the variation of the stationary metric, $h_{ij} = 2\delta M R^{-1} \delta_{ij} + \mathcal{O}(r^{-2})$. Using the above expressions in eq. (7.37) yields the desired result (7.34). $\square$

In order to conclude the derivation of the first law, we note that the integral in the variation formula (7.34) can be evaluated with the help of the Einstein–Hilbert action principle,

$$\frac{1}{\sqrt{-g}} \, \delta(R\sqrt{-g}) = -G^{\mu\nu} h_{\mu\nu} + 2\nabla_\nu \nabla^{[\mu} h^{\nu]}_{\mu}, \tag{7.39}$$

where $-\nabla_\nu \nabla^{[\mu} h^{\nu]}_{\mu} = -\nabla_\nu h^\nu = d^\dagger h$. Using $\delta k^\mu = 0$ and $*k = i_k \eta$ we have $\delta(\frac{1}{\sqrt{-g}} * k) = 0$ and thus $\delta(R * k) = \frac{1}{\sqrt{-g}} \delta(R\sqrt{-g}) * k$. Equation (7.39) now assumes the form

$$d^\dagger h * k = -\frac{1}{2} \left[G^{\mu\nu} h_{\mu\nu} * k + \delta(R * k) \right]. \tag{7.40}$$

The desired mass variation formula is finally obtained from varying the mass formula (7.17) and using eqs. (7.34) and (7.40) to substitute $A\delta\kappa + 8\pi J_H \delta\Omega_H$:

$$
\begin{aligned}
4\pi \, \delta M &= \kappa \, \delta\mathcal{A} + 8\pi \, \Omega_H \delta J_H + \mathcal{A}\delta\kappa + 8\pi J_H \delta\Omega_H - \delta \int_\Sigma *R(k) \\
&= \kappa \, \delta\mathcal{A} + 8\pi \, \Omega_H \, \delta J_H - 4\pi \, \delta M \\
&\quad + \frac{1}{2} \int_\Sigma G^{\mu\nu} h_{\mu\nu} * k + \frac{1}{2}\delta \int_\Sigma R * k - \delta \int_\Sigma *R(k).
\end{aligned}
$$

In terms of the 1–form $G(k)$ (with components $G(k)_\mu = G_{\mu\nu} k^\nu$) we also have $*R(k) - \frac{1}{2} R * k = *G(k)$. This yields the final result:

Theorem 7.8 *(first law) The variation of the total mass between two infinitesimally neighboring stationary (and axisymmetric) black hole solutions can be expressed in terms of the horizon quantities κ, $\delta\mathcal{A}$, Ω_H and δJ_H, the Einstein tensor and the asymptotically timelike Killing field k by*

$$\delta M = \frac{1}{8\pi}\kappa \, \delta\mathcal{A} + \Omega_H \, \delta J_H + \frac{1}{16\pi} \int_\Sigma G^{\mu\nu} \delta g_{\mu\nu} * k - \frac{1}{8\pi}\delta \int_\Sigma *G(k). \tag{7.41}$$

For a large class of field theories it is possible to express the integrals in eq. (7.41) in terms of the matter Lagrangian $\mathcal{L}$: If $\mathcal{L}$ does not depend on derivatives of the metric then the variation of the matter action with respect to $g_{\mu\nu}$ does not lead to integrations by parts. In this case the stress–energy tensor is obtained from

$$\delta_g(\mathcal{L}\eta) = -\frac{1}{2} T^{\mu\nu} \delta g_{\mu\nu} \, \eta, \qquad T_{\mu\nu} = 2\frac{\partial \mathcal{L}}{\partial g^{\mu\nu}} - \mathcal{L} g_{\mu\nu}, \tag{7.42}$$

where δ_g denotes variations with respect to the metric and where the first equation remains valid if η is replaced by $*k = i_k\eta$. Introducing the tensor $(\mathcal{L}_g)_{\mu\nu} \equiv \partial\mathcal{L}/\partial g^{\mu\nu}$ and the 1–form $(\mathcal{L}_g(k))_\mu \equiv$

$(\mathcal{L}_g)_{\mu\nu}k^{\nu}$, Einstein's equations imply

$$\delta_g \left(\mathcal{L} * k\right) = -\frac{1}{16\pi} G^{\mu\nu} \delta g_{\mu\nu} * k, \qquad 2\mathcal{L}_g(k) - \mathcal{L}k = \frac{1}{8\pi} G(k).$$

Using these relations in the mass formula (7.17) and in the variation expression (7.41), and denoting the variation with respect to the matter fields with δ_χ finally yields

$$M - \frac{1}{4\pi}\kappa\mathcal{A} - 2\Omega_H J_H = -4I_1 - 2I_2 + 2I_3, \qquad (7.43)$$

$$\delta M - \frac{1}{8\pi}\kappa\delta\mathcal{A} - \Omega_H\delta J_H = -2\delta I_1 + \delta_\chi I_2, \qquad (7.44)$$

where the integrals I_1, I_2 and I_3 are given in terms of the matter Lagrangian $\mathcal{L}$ and the asymptotically timelike Killing field k:

$$I_1 = \int_\Sigma *\mathcal{L}_g(k), \qquad I_2 = \int_\Sigma \mathcal{L}*k, \qquad I_3 = \int_\Sigma \mathrm{tr}(\mathcal{L}_g)*k. \quad (7.45)$$

For any Lagrangian field theory where the dynamical fields are stationary for stationary physical solutions, the integrals in the variation formula (7.44) can be converted into surface terms (see Wald 1993b). In general this is, however, not true for the integrals appearing in the mass formula (7.43) itself (see Heusler and Straumann 1993a, 1993b).

As is shown in Brodbeck *et al.* (1996), the mass variation formula can be obtained from a very simple and short calculation in the spherically symmetric case. In particular, the spherically symmetric version applies to a variety of recently found black hole configurations "with hair", such as self–gravitating Yang–Mills solutions (Volkov and Gal'tsov 1989, Bizon 1990, Künzle and Masood–ul–Alam 1990), Skyrme fields (Droz *et al.* 1991), dilaton fields (Lavrelashvili and Maison 1993), or Yang–Mills–Higgs configurations (Breitenlohner *et al.* 1992, Greene *et al.* 1993).

As an example, we evaluate the expressions (7.43) and (7.44) for nonrotating, electrovac black holes. Using the Maxwell Lagrangian $\mathcal{L} = (16\pi)^{-1}F_{\alpha\beta}F_{\gamma\delta}g^{\alpha\gamma}g^{\beta\delta}$ and the electric and magnetic fields $E \equiv E_k = -i_k F$, $B \equiv B_k = i_k * F$ (see section 5.1) we find

$$*\mathcal{L}_g(k) = -\frac{1}{8\pi}\left(E \wedge *F\right),$$

$$\mathrm{tr}\left(\mathcal{L}_g\right)*k = 2\mathcal{L}*k = \frac{1}{4\pi}\left(B \wedge F - E \wedge *F\right).$$

By virtue of Maxwell's equations, $dF = d * F = 0$, $E = d\phi$ and $B = d\psi$, the integrand on the r.h.s. of the mass formula (7.43) becomes an exact differential form:

$$-4I_1 - 2I_2 + 2I_3 = \frac{1}{4\pi} \int_\Sigma (B \wedge F + E \wedge *F)$$

$$= \frac{1}{4\pi} \int_\Sigma d(\psi F + \phi * F) = \frac{1}{4\pi} \int_{\partial\Sigma} (\psi F + \phi * F).$$

In the next chapter we shall see that both the electric and the magnetic potential assume constant values on the horizon. Adopting a gauge where ϕ and ψ vanish asymptotically, one obtains (using $Q = -(4\pi)^{-1} \int_{S_\infty^2} *F = -(4\pi)^{-1} \int_{\mathcal{H}} *F$ and $P = -(4\pi)^{-1} \int_{\mathcal{H}} F$) the mass formula

$$M = \frac{1}{4\pi} \kappa \mathcal{A} + \phi_H Q + \psi_H P. \qquad (7.46)$$

In order to evaluate the mass variation formula, we must compute $-2\delta I_1 + \delta_A I_2$. First, we have

$$\delta_A I_2 = \delta_A \frac{1}{8\pi} \int_\Sigma (F|F) * k = \frac{1}{4\pi} \int_\Sigma i_k (\delta F \wedge *F)$$

$$= -\frac{1}{4\pi} \int_\Sigma (\delta E \wedge *F - \delta F \wedge B)$$

$$= -\frac{1}{4\pi} \int_{\delta\Sigma} \delta\phi * F + \frac{1}{4\pi} \int_{\delta\Sigma} \delta A \wedge B,$$

where we have again used $E = d\phi$ and $dF = d * F = dB = 0$. The variation of I_1 becomes

$$-2\delta I_1 = \frac{1}{4\pi} \int_{\delta\Sigma} \delta\phi * F + \frac{1}{4\pi} \int_{\delta\Sigma} \phi\delta * F.$$

Hence, the terms involving variations of the electric potentials cancel each other. Again using the formula for the total charge yields

$$-2\delta I_1 + \delta_A I_2 = \phi_H \delta Q + \frac{1}{4\pi} \int_{\delta\Sigma} \delta A \wedge B.$$

It remains to compute the last boundary integral. Taking advantage of the fact that $\delta A \wedge B = \delta A \wedge d\psi = -d(\psi\delta A) - \psi\delta F$, we obtain $(4\pi)^{-1} \int_{\delta\Sigma} \delta A \wedge B = \psi\delta P$. This finally yields the desired variation formula

$$\delta M = \frac{1}{8\pi} \kappa \delta\mathcal{A} + \phi_H \delta Q + \psi_H \delta P. \qquad (7.47)$$

It is an easy exercise to verify these formulae for the Reissner-Nordström solution, for which $(w \equiv \sqrt{M^2 - Q^2})$

$$\kappa = \frac{w}{(M+w)^2}, \quad \frac{\mathcal{A}}{4\pi} = (M+w)^2, \quad \phi_H = \frac{Q}{M+w}.$$

To conclude this section, we note that the vacuum variation formula for the Kerr solution is immediately obtained by differentiating the explicit expression

$$\frac{\mathcal{A}}{8\pi} = M^2 + \sqrt{M^4 - J_H^2} \qquad (7.48)$$

for the area of the Kerr black hole in terms of M and J_H. Using also the formulae

$$\kappa = \frac{1}{2M} \frac{\sqrt{M^4 - J_H^2}}{M^2 + \sqrt{M^4 - J_H^2}}, \quad \Omega_H = \kappa \frac{J_H}{\sqrt{M^4 - J_H^2}}, \qquad (7.49)$$

one finds

$$\frac{1}{8\pi} \delta\mathcal{A} = \frac{1}{\kappa} \delta M - \frac{1}{\kappa} \Omega_H \delta J_H, \qquad (7.50)$$

in agreement with the general vacuum expression (7.41). This is, in fact, the way Bekenstein (1973) originally derived the mass variation formula for the Kerr metric.

7.3 The second law

In the previous sections we derived the zeroth and first laws of black hole physics. These are based on the fact that the event horizon of a stationary black hole is a Killing horizon. As already mentioned, the proof of this theorem makes use of local techniques which do not require the knowledge of the global evolution of spacetime. These concepts - notably the notion of trapped surfaces (Penrose 1965b) and the kinematics of geodesic congruences - which also lead to the famous singularity theorems, are also used to derive the area increase theorem. This theorem, known as the second law of black hole physics, applies to the event horizon of a strongly asymptotically predictable spacetime. Here we give only a brief outline of its proof and refer the reader to the literature for more detailed presentations (see, e.g., Wald 1984, chap. 12).

Our starting point is the generalized Raychaudhuri (1955) equation for a *null* geodesic congruence generated by ξ^μ, say. It is most easely derived in the complex null tetrad of the null vectors ξ and η, $(\xi|\eta) = -1$, and the pair of complex conjugate vectors $v_\pm = (v_1 \pm iv_2)/\sqrt{2}$, where v_1 and v_2 are real spacelike vectors orthonormal to ξ and η (Newman and Penrose 1962; see Wald 1984 or DeFelice and Clarke 1990 for derivations). Consider the expansion $\theta = \nabla_\mu \xi^\mu$ of the geodesic *null* congruence with tangent vector ξ^μ with respect to an *affine* parameter λ. The rate of change of θ is then given by

$$\frac{d\theta}{d\lambda} = -\frac{1}{2}\theta^2 - \sigma_{\mu\nu}\sigma^{\mu\nu} + \omega_{\mu\nu}\omega^{\mu\nu} - R(\xi,\xi), \qquad (7.51)$$

where $\sigma_{\mu\nu}$ and $\omega_{\mu\nu}$ are the shear and the twist of the congruence, respectively. (Recall that the factor in front of θ^2 is $\frac{1}{3}$ in the case of a timelike congruence.)

The event horizon, $H^+[\xi]$, is a null hypersurface with vanishing twist. Since θ corresponds to the relative change of an area element when Lie transported along a generator of this null surface, we have

$$-\left[\frac{d\theta}{d\lambda} + \frac{1}{2}\theta^2\right] = \frac{1}{2}\sigma^2 + R(\xi,\xi), \quad \text{where } \theta = \mathcal{A}^{-1}\frac{d\mathcal{A}}{d\lambda}. \quad (7.52)$$

The r.h.s. of this equation is manifestly non–negative, provided that $R(\xi,\xi) \geq 0$. By virtue of Einstein's equations and the fact that ξ is null, this is equivalent to the null energy condition $T(\xi,\xi) \geq 0$. Assuming that the stress–energy tensor is subject to this condition, we can integrate the resulting inequality, which yields

$$\frac{1}{\theta} \geq \frac{1}{\theta(\lambda_0)} + \frac{\lambda - \lambda_0}{2}. \qquad (7.53)$$

This shows that the expansion rate diverges at a finite affine length $\lambda_\star$ if the congruence is initially converging, i.e., if $\theta(\lambda_0) \leq 0$. One thus has the following theorem (see Wald 1984):

Theorem 7.9 *Let θ be the expansion of a hypersurface orthogonal null geodesic congruence and let Einstein's equations hold with matter satisfying the weak or the strong energy condition. If $\theta(\lambda_0)$ is negative at any point on any geodesic in the congruence, then $\theta(\lambda_\star) = -\infty$ on that geodesic for a finite affine length $\lambda_\star$ with $\lambda_\star \leq \lambda_0 + 2|\theta(\lambda_0)^{-1}|$.*

This theorem is now used to conclude that the expansion cannot be negative on the event horizon of a predictable black hole. As in the proof of the zeroth law, the reasoning is less difficult if the null–generators of the horizon are required to be geodesically complete: For, if θ were negative at a point on $H^+[\xi]$, then the corresponding null–generator would reach a point of infinite convergence after a finite affine distance, beyond which the generators intersect and must leave $H^+[\xi]$. This is, however, in contradiction to the fact that the null–generators of the horizon have no future endpoints. Using global hyperbolicity, a more involved argument can be applied to conclude that θ cannot become negative on the horizon, even without assuming completeness of the null–generators (see Hawking and Ellis 1973, Carter 1979 and Wald 1984 for comprehensive discussions). Once $\theta \geq 0$ on $H^+[\xi]$ is established, the second law of black hole dynamics follows immediately:

Theorem 7.10 *(second law) Let (M,g) be a strongly asymptotically predictable spacetime with spacelike Cauchy surfaces Σ_1 and Σ_2 for the globally hyperbolic region $\check{V}$. Let $\mathcal{H}_1 = H^+ \cap \Sigma_1$ and $\mathcal{H}_2 = H^+ \cap \Sigma_2$ be the event horizons at time Σ_1 and Σ_2, respectively, and let $\mathcal{A}_i = \mathcal{A}(\mathcal{H}_i)$ denote the area of $\mathcal{H}_i$. If Σ_2 is contained in the chronological future of Σ_1, $\Sigma_2 \subset I^+(\Sigma_1)$, then $\mathcal{A}_2 \geq \mathcal{A}_1$.*

Proof Since each point $p_1 \in \mathcal{H}_1$ lies on a future inextensible null geodesic of H which must intersect the Cauchy surface Σ_2 at $p_2 \in \mathcal{H}_2$, say, one obtains a map $f\colon \mathcal{H}_1 \to \mathcal{H}_2$ (with $\mathcal{H}_2 \supseteq f(\mathcal{H}_1)$). Since the expansion is non–negative, this implies that $\mathcal{A}(\mathcal{H}_2) \geq \mathcal{A}(f(\mathcal{H}_1)) \geq \mathcal{A}(\mathcal{H}_1)$. $\qquad\qquad\square$

7.4 The generalized entropy

The analogy between the surface gravity and the thermodynamic temperature suggests that it is impossible to achieve a black hole state with vanishing surface gravity. In the Kerr geometry, this would correspond to an extreme solution with $M^4 = J_H^2$. The impossibility of reaching this limit by a finite sequence of operations (Wald 1974, Israel 1986), resembles the third law of ordinary thermodynamics. However, the third law of black hole physics stands on a less solid foundation than the other three laws, which are rigorous theorems in differential geometry.

The formal relationship between the laws of thermodynamics and those of black hole physics suggests the identifications $E \doteq M$, $T \doteq \alpha \kappa$ and $S \doteq \alpha^{-1} \frac{A}{8\pi}$, α being an arbitrary constant within the classical framework. Although the energy E and the total mass M actually do represent the same physical quantity, the analogy remains on a formal level, since the thermodynamic temperature of a black hole (being a perfect absorber and emitting nothing) is absolute zero in *classical* relativity.

The situation changes if the prediction due to Hawking (1975) is taken into account (see also Wald 1975): This states that a black hole emits particles with a black–body spectrum corresponding to a temperature of $T = \kappa/2\pi$ $(= \frac{\kappa}{2\pi}\hbar)$. This immediately suggests that $\mathcal{A}/4$ $(= \frac{\mathcal{A}}{4}\frac{kc^3}{G\hbar})$ *is* the entropy of a black hole. As a matter of fact, Gibbons and Hawking (1977) gave strong evidence for the validity of this formula, using the Euclidean approach to quantum gravity. More recently, Brown and York (1993a, 1993b) have also given a derivation which is based on the microcanonical ensemble, rather than the (modified) canonical one as used in earlier attempts.

More support for the conjecture that $\mathcal{A}/4$ must be considered as the entropy of a black hole is provided by the following consideration: Both the area theorem and the second law of ordinary thermodynamics can be violated by taking quantum processes into account. For instance, the area of a black hole evaporating due to Hawking radiation decreases to zero. On the other hand, the thermodynamic entropy outside the horizon may be decreased by dropping matter into the black hole. However, both problems may be solved simultaneously by introducing the *generalized entropy*

$$S_{\text{gen}} = S + \frac{\mathcal{A}}{4}, \qquad (7.54)$$

and conjecturing that

$$\delta S_{\text{gen}} \geq 0 \qquad (7.55)$$

in any process (Bekenstein 1974a). The generalized version of the second law therefore states that in any process of the above kind a decrease in S is accompanied by an increase in the area $\mathcal{A}$ (and vice versa), such that $\delta S_{\text{gen}} \geq 0$ remains valid. If this turns out to be correct, the laws of black hole thermodynamics may be considered to be the ordinary laws of thermodynamics for a self–gravitating quantum system containing a black hole.

In conclusion, we would like to emphasize that the laws of black hole mechanics are still subject to numerous investigations. On the one hand, one would like to extend their validity to the largest possible class of matter models. Of particular interest are, for instance, Yang–Mills(–Higgs) theories or models with Skyrme fields, both of which are not covered by the uniqueness theorems. On the other hand, recent investigations have shown that the first law applies to arbitrary theories of gravity which can be derived from a diffeomorphism–covariant Lagrangian. Moreover, the new derivations of the first law have also led to a deeper understanding of the entropy of a black hole. In fact, the latter turns out to be the Noether charge which is associated with the symmetry generated by the horizon Killing field (Wald 1993b, Iyer and Wald 1994). The reader is referred to Wald (1994) for an up–to–date presentation of the status of black hole thermodynamics and to Davies (1978), Jacobson and Kang (1993), Hayward (1994c) and Jacobson *et al.* (1994) for more information.

8

Integrability and divergence identities

The strong rigidity theorem implies that stationary black hole spacetimes are either axisymmetric or have a nonrotating horizon. The uniqueness theorems are, however, based on stronger assumptions: In the nonrotating case, staticity is required whereas the uniqueness of the Kerr–Newman family is established for circular spacetimes. The first purpose of this chapter is therefore to discuss the circumstances under which the integrability conditions can be established.

Our second aim is to present a systematic approach to divergence identities for spacetimes with one Killing field. In particular, we consider the stationary Einstein–Maxwell equations and derive a mass formula for nonrotating - not necessarily static - electrovac black hole spacetimes.

The chapter is organized as follows: In the first section we recall that the two Killing fields in a stationary and axisymmetric domain fulfil the integrability conditions if the Ricci–circularity conditions hold. As an application, we establish the circularity theorem for electrovac spacetimes.

The second section is devoted to the staticity theorem. As mentioned earlier, the staticity issue is considerably more involved than the circularity problem. The original proof of the staticity theorem for black hole spacetimes applied to the vacuum case (Hawking and Ellis 1973). Here we present a different proof which establishes the equivalence of staticity and Ricci–staticity for a strictly stationary domain. Since our reasoning involves no potentials, it is valid under less restrictive topological conditions. We conclude this section with some comments on the electrovac staticity theorem, which is still subject to investigations. Sudarsky and Wald (1993) were recently able to settle the problem for spacetimes admitting a foliation by maximal slices. The existence of the latter was meanwhile established by Chruściel and Wald (1994a) (see also Bartnik 1984).

Maxwell's equations and the Einstein equations for the Ricci 1–form (with respect to the stationary Killing field) yield four expressions for the Laplacians of the electric, the magnetic, the gravitational and the combined twist potentials. Using Stokes' theorem, we shall gain four divergence identities from these relations in the third section. These identities are then used to derive the formula $M^2 = (\frac{\kappa}{4\pi}\mathcal{A})^2 + Q^2 + P^2$ for a nonrotating - not necessarily static - stationary black hole spacetime (with total mass M, electric charge Q, magnetic charge P, surface gravity κ and horizon area $\mathcal{A}$). In particular, this yields an elementary proof of the Bogomol'nyi inequality $M \geq |Q|$ (see Gibbons and Hull 1982, Gibbons *et al.* 1983) for a stationary electrovac black hole spacetime with nonrotating horizon.

Further applications of the divergence formulae are discussed in the last section. Restricting ourselves to the static case, we recover the fact that the electric and the magnetic potentials depend only on the norm of the Killing field. This crucial result was the key to the extension of the vacuum uniqueness theorem (Israel 1967) to electrovac configurations (Israel 1968, Müller zum Hagen *et al.* 1974). We also re–derive the identity used in Carter's (1987) attempt to prove the electrovac staticity theorem. Eventually, we discuss the relation between the new potentials introduced in this chapter and the Ernst potentials E and Λ.

Some derivations and results discussed in this chapter are, to the best of our knowledge, not available elsewhere. In particular, the version of the staticity theorem presented in the second section and the derivation of the mass formula given in the third section seem to be new.

8.1 The circularity theorem

The strong rigidity theorem implies that the null–generator of the horizon of a stationary black hole either coincides with the asymptotically timelike Killing field k, or the spacetime admits a further, axisymmetric Killing field m. In view of numerous applications, in particular of the uniqueness theorems, one would like to write the metric in the second case in the Papapetrou form discussed in chapter 3. This can be achieved by establishing the Frobenius integrability conditions, $(\omega_k|m) = (\omega_m|k) = 0$. As we showed in theorem 2.20, these conditions are fulfilled if the Ricci–circularity conditions, $m \wedge k \wedge R(k) = k \wedge m \wedge R(m) = 0$, hold. The key point was the observation that the latter relations are actually the exterior derivatives of the former, $d(\omega_k|m) = - * (m \wedge k \wedge R(k))$. In addition, we used the fact that $d(\omega_k|m) = 0$ implies $(\omega_k|m) = 0$ in every domain of spacetime intersecting the rotation axis.

In combination with Einstein's equations, the above connection provides the link between the integrability conditions on the one hand and the symmetry properties of the matter fields on the other:

Corollary 8.1 *Consider an asymptotically flat, stationary and axisymmetric domain of outer communications. Then the latter is circular if and only if it is Ricci–circular or, equivalently, if and only if the stress–energy tensor fulfils the conditions*

$$m \wedge k \wedge T(k) = 0, \quad k \wedge m \wedge T(m) = 0. \qquad (8.1)$$

The above condition is trivially satisfied for vacuum spacetimes. In the case of a stationary and axisymmetric electromagnetic field ($L_k F = L_m F = 0$), Maxwell's equations imply $F(k,m) = 0$ and $(*F)(k,m) = 0$ (see proposition 5.6), which establishes Ricci–circularity (see eqs. (5.52), (5.53)). Hence, for rotating electrovac black hole solutions the domain of outer communications is circular. A similar argument establishes circularity in the presence of other matter models, such as scalar fields or, more generally, mappings from spacetime into Riemannian target manifolds (see section 12.1). The theorem does, however, not generalize to non–Abelian gauge fields in a straightforward way.

8.2 The staticity theorem

Let us now consider the situation where spacetime admits only the asymptotically timelike Killing field k. Then the rigidity theorem states that k coincides with the null–generator of the Killing horizon. In this case, one would wish to establish staticity which, as before, involves establishing the appropriate integrability conditions.

The situation is, however, more complicated than in the case of two Killing fields. First, one has to establish *strict* stationarity, that is, one would like to exclude ergoregions in the nonrotating case. This problem, first discussed by Hajicek (1973, 1975) and Hawking and Ellis (1973), was solved only recently by Sudarsky and Wald (1992, 1993), assuming a foliation by maximal slices (Chruściel 1994a). Once ergoregions are excluded, it still remains to establish staticity (see, e.g., Carter 1987 for a detailed discussion). This was originally done by Lichnerowicz (1955) and Choquet–Bruhat for spacetimes which are topologically $I\!R^4$. Later, Hawking (1972) was able to extend the theorem to the vacuum *black hole* case. An early success of Carter's (1973a) for electrovac spacetimes was, unfortunately, due to a sign error. The latter was cured by requiring a rather "unphysical" inequality between the norm of the Killing field and the electric potential (Carter 1987). As already mentioned, Sudarsky and Wald (1992, 1993) were eventually able to derive staticity for electrovac black holes by using the generalized version of the first law of black hole physics.

Here we present a version of the staticity theorem which establishes the equivalence of Ricci–staticity and metric staticity in the strictly stationary case. The theorem is stronger than the vacuum version given in Hawking and Ellis (1973), since it applies to arbitrary matter fields satisfying the Ricci condition $k \wedge T(k) = 0$. In addition, it does not involve the twist potential. However, the theorem does not solve the electrovac staticity problem (unless the Poynting vector vanishes) since the symmetry conditions and Maxwell's equations do not imply Ricci–staticity. (This is, in fact, the reason why the electrovac staticity theorem is much harder to establish than the corresponding circularity problem.) In all that follows we use the general identities of chapter 2 for the stationary Killing field, i.e., for $K \equiv k$, $N \equiv -V$ and $\omega \equiv \omega_k$.

Theorem 8.2 *Consider an asymptotically flat spacetime with a nondegenerate, nonrotating (not necessarily connected) Killing horizon and strictly stationary domain of outer communications. Then the latter is static if and only if it is Ricci-static or, equivalently, if and only if the stress–energy tensor fulfils the condition*

$$k \wedge T(k) = 0 \,. \tag{8.2}$$

Proof Let us first prove the trivial direction, establishing that staticity implies $k \wedge T(k) = 0$. This is an immediate consequence of the identity (2.16) and Einstein's equations:

$$d\omega = 8\pi G * (k \wedge T(k)). \qquad (8.3)$$

Let us now consider the converse problem. This obviously requires global arguments, since we have to show that $d\omega = 0$ implies $\omega = 0$. The proof makes essential use of the Killing property of the horizon. The latter, however, is not required to be connected. The starting point is the *identity*

$$d\left(\omega \wedge \frac{k}{V}\right) = d\omega \wedge \frac{k}{V} - 2\frac{(\omega|\omega)}{V^2} * k, \qquad (8.4)$$

which is obtained from eq. (2.12) and the fact that $i_k\omega = 0$ implies $\omega \wedge *(k \wedge \omega) = -\omega \wedge i_k * \omega = i_k(\omega \wedge *\omega)$. If spacetime is Ricci–static, that is, if $d\omega = 0$, then Stokes' theorem yields

$$\int_{\partial\Sigma} \omega \wedge \frac{k}{V} = -2\int_\Sigma \frac{(\omega|\omega)}{V^2} i_k\eta, \qquad (8.5)$$

where Σ denotes the 3–dimensional spacelike hypersurface extending from its inner boundary, $\mathcal{H} = \Sigma \cap H[k]$, to spacelike infinity, and $i_k\eta = *k$ is the volume 3–form on Σ. In order to conclude that ω vanishes, we have to argue that the boundary terms on the l.h.s. of eq. (8.5) vanish. This is clearly the case for the contribution from S_∞^2, since $V \to 1$ and $\omega \to 0$. However, since V vanishes on each component $\mathcal{H}_i$ of the horizon, it is not obvious that $\int_{\mathcal{H}_i} \omega \wedge \frac{k}{V}$ vanishes as well. In order to see this, we note that

$$\left(\frac{k \wedge \omega}{V} \middle| \frac{k \wedge \omega}{V}\right) = \frac{1}{V^2}(k|k)(\omega|\omega) = -\frac{(\omega|\omega)}{V} = 2\kappa^2$$

remains finite on the horizon. (Here we have used eqs. (2.11), (6.14) and (6.16) in the last step.) Using also eq. (7.15) to evaluate $*\Lambda = (k \wedge \omega)/V$ on the horizon yields the desired result,

$$\int_{\mathcal{H}_i} \frac{k \wedge \omega}{V} = -\int_{\mathcal{H}_i} i_n i_k * \left(\frac{k \wedge \omega}{V}\right) d\mathcal{A} = \int_{\mathcal{H}_i} i_n * \left(k \wedge \frac{k \wedge \omega}{V}\right) d\mathcal{A}, \qquad (8.6)$$

where n is the second future–directed null vector orthogonal to the horizon, normalized such that $(k|n) = -1$. Hence, the volume integral on the r.h.s. of eq. (8.5) must vanish. Since k is orthogonal to ω and k is nowhere spacelike in the domain of outer communications, ω is nowhere timelike. The integrand is therefore semi–definite and the integral can vanish only if $\omega = 0$. □

As already mentioned, no restrictions concerning the topology of $\mathcal{H}$ enter into the above reasoning. A closely related, but less powerful, proof is obtained by using $d\omega = 0$ to introduce the potential U with $\omega = dU$. Taking advantage of Stokes' theorem (2.8) for the 1–form $\alpha = U\omega/V^2$ and using the identity $d^\dagger(\omega/V^2) = 0$, as well as the fact that U assumes a constant value on the horizon, also yields the conclusion $\omega = 0$. However, the existence of the potential U depends, of course, on stronger topological assumptions.

As an immediate consequence of the above theorem, we are also able to draw the conclusion that there must exist an ergoregion if the domain of outer communications is not static, but $k \wedge T(k) = 0$. In fact, it is easy to see that the converse statement holds as well (Hawking and Ellis 1973). Hence, strict stationarity and staticity imply each other if the stress–energy tensor fulfils $k \wedge T(k) = 0$.

Let us conclude this section by applying the above theorem to the electrovac case. In section 5.1 we derived the identity $d\omega = 8\pi G * (k \wedge T(k)) = 2G\,B_k \wedge E_k$ (see eq. (5.13) with $K \equiv k$), where B_k and E_k are the magnetic and the electric components of the field tensor F with respect to the Killing field k: $B_k = i_k * F$, $E_k = -i_k F$. From this and definition (5.10) we conclude that Ricci–staticity is equivalent to the vanishing of the Poynting vector, since

$$S_k = -\frac{1}{2G} * (k \wedge d\omega). \tag{8.7}$$

(Note that $d\omega = 0 \iff k \wedge d\omega = 0$ if $V \neq 0$.) This yields the following corollary to the above staticity theorem:

Corollary 8.3 *Consider an asymptotically flat electrovac space-time with nondegenerate, nonrotating (not necessarily connected) Killing horizon and strictly stationary domain of outer communications. Then the latter is static if and only if the Poynting vector associated with the Killing trajectories vanishes or, equivalently, if and only if the electric and the magnetic field are parallel, or one of them vanishes.*

In view of applications to other matter models, we would like to emphasize again that the strong rigidity theorem does *not* imply that the domain of outer communications is either static or circular, as is sometimes claimed. In order to establish these properties one has to perform a number of intermediate steps. In the nonrotating case, these steps are not even trivial for vacuum configurations and, if matter fields are involved, make essential use of their symmetry properties. We shall return to this problem when discussing the extension of the uniqueness theorems to other nonlinear classical field theories, such as self–gravitating harmonic mappings.

8.3 Divergence identities

Divergence identities play a central role in the original proof of the uniqueness theorem for nonrotating black holes. Although we shall mainly be concerned with the modern approach to the theorem (Bunting and Masood–ul–Alam 1987), based on the positive energy theorem (Schoen and Yau 1979, 1981, Witten 1981), we devote this section to a systematic investigation of divergence formulae for the *stationary* Einstein–Maxwell system. This generalizes the methods invented by Israel (1968) and Müller zum Hagen *et al.* (1974) for the corresponding *static* equations. (We also refer to Lindblom (1980) for a systematic derivation of divergeence formulae.)

As a first application, we establish the relation $M^2 = (\frac{\kappa}{4\pi}\mathcal{A})^2 + Q^2 + P^2$ for a nonrotating, not necessarily static, electrovac black hole spacetime. This also proves the inequality $M \geq |Q|$, where (for vanishing magnetic monopole moment P) equality holds if and only if the horizon is degenerate.

The methods presented in this section might also turn out to be of some relevance for the uniqueness proof of the Papapetrou–Majumdar multi–black hole solution (Papapetrou 1945, Majumdar 1947) and the Israel–Wilson (1972) metric. We shall, however, not treat these issues here and refer the reader to Chruściel and Nadirashvili (1995) and Heusler (1995c) for recent results.

Throughout the remainder of this chapter we assume that the domain of outer communications is simply connected. We use the equations derived in sections 5.1 and 5.2 for the asymptotically timelike Killing field, that is, for $K \equiv k$, $N \equiv -V$, $\omega \equiv \omega_k$, $E \equiv E_k$ and $B \equiv B_k$. The source–free Maxwell equations (5.25),

(5.26) for a stationary electromagnetic field now imply

$$d^\dagger\left(\frac{E}{V}\right) = -2\frac{(\omega|B)}{V^2}, \qquad d^\dagger\left(\frac{B}{V}\right) = 2\frac{(\omega|E)}{V^2}, \qquad (8.8)$$

where $E = d\phi$, $B = d\psi$. In addition, we have the general Killing field identities (2.16)-(2.18), where Einstein's equations imply that $R(k,k) = (E|E) + (B|B)$ and $*(k \wedge R(k)) = -2\,E \wedge B$ (see eqs. (5.12), (5.13)). Hence, the four potentials ϕ, ψ, V and U fulfil the following (Poisson) equations:

$$d^\dagger\left(\frac{d\phi}{V}\right) = -2\frac{(\omega|d\psi)}{V^2}, \qquad d^\dagger\left(\frac{d\psi}{V}\right) = 2\frac{(\omega|d\phi)}{V^2}, \quad (8.9)$$

$$d^\dagger\left(\frac{dV}{V}\right) = 4\frac{(\omega|\omega)}{V^2} - 2\frac{(d\phi|d\phi) + (d\psi|d\psi)}{V}, \qquad (8.10)$$

$$d^\dagger\left(\frac{\omega}{V^2}\right) = 0, \quad \text{where} \quad \omega = dU + \psi d\phi - \phi d\psi. \quad (8.11)$$

Before we apply Stokes' theorem to these equations, we first prove the following proposition:

Proposition 8.4 *Let ϕ and ψ be the electric and the magnetic potentials, defined with respect to the stationary Killing field k, i.e., $d\phi = -i_k F$, $d\psi = i_k * F$. Let $V = -(k|k)$ and let U be the combined twist potential, $dU = \omega - \psi d\phi + \phi d\psi$. Then ϕ, ψ, V and U are invariant under the action of k and assume constant values on the horizon.*

Proof The Lie derivative of a function is $L_k = i_k d$. Hence

$$L_k \psi = i_k\, d\psi = i_k\, B = i_k\, i_k * F = 0,$$

and similarly for the other potentials, which proves the first assertion.

By definition, V vanishes on the Killing horizon $H[k]$. In order to show that the other potentials assume constant values as well, we first recall proposition 6.15, according to which $R(k,k) = 0$ on $H[k]$. The Einstein equation $R(k,k) = (E|E) + (B|B)$ now implies that both E and B are null on $H[k]$. This is due to the fact that they are orthogonal to k (by definition) and k is null on the horizon. Hence, E and B are also proportional to k on $H[k]$,

$$E = \sigma_E\, k, \qquad B = \sigma_B\, k \qquad \text{on } H[k]. \qquad (8.12)$$

For an arbitrary vector field Z tangent to the horizon we now have
$$L_Z \phi = i_Z d\phi = i_Z E = \sigma_E (Z|k) = 0 \quad \text{on } H[k],$$
and similarly for ψ. Using the definition of the potential U and the general property $\omega = 0$ of a Killing horizon, we finally obtain
$$L_Z U = i_Z (\omega - \psi E + \phi B) = (Z|\omega) = 0 \quad \text{on } H[k],$$
which completes the proof.   □

We now set up the Komar formulae for the total mass M and the charges Q and P. In terms of the 2–forms $*dk$, $*F$ and F, these quantities are defined by the flux integrals

$$M, M_H = -\frac{1}{8\pi} \int_{S^2_\infty, \mathcal{H}} *dk, \tag{8.13}$$

$$Q, Q_H = -\frac{1}{4\pi} \int_{S^2_\infty, \mathcal{H}} *F, \tag{8.14}$$

$$P, P_H = -\frac{1}{4\pi} \int_{S^2_\infty, \mathcal{H}} F. \tag{8.15}$$

Here we have also introduced the quantities M_H, Q_H and P_H, which are defined by the corresponding Komar integrals over $\mathcal{H}$ instead of S^2_∞. (Note also that $M_H = (4\pi)^{-1}\kappa\mathcal{A}$, as is seen from eq. (7.16) in the nonrotating case.) The following proposition gives an alternative set of formulae for the above quantities:

Proposition 8.5 *Consider a stationary electrovac black hole spacetime with asymptotically timelike Killing field k and non-rotating horizon $H[k]$. Then the total "charges" M, Q and P and the corresponding horizon quantities M_H, Q_H and P_H defined in eqs. (8.13)-(8.15) are given by*

$$M, M_H = \frac{1}{8\pi} \int_{S^2_\infty, \mathcal{H}} * \left(k \wedge \frac{dV}{V} \right), \tag{8.16}$$

$$Q, Q_H = -\frac{1}{4\pi} \int_{S^2_\infty, \mathcal{H}} * \left(k \wedge \frac{E}{V} \right), \tag{8.17}$$

$$P, P_H = -\frac{1}{4\pi} \int_{S^2_\infty, \mathcal{H}} * \left(k \wedge \frac{B}{V} \right). \tag{8.18}$$

Proof Using the identities (2.10) and (5.7) for dk and F,
$$*dk = -\frac{2}{V}(k \wedge \omega) - * \left(k \wedge \frac{dV}{V} \right), \qquad F = \frac{k \wedge E}{V} + * \frac{k \wedge B}{V},$$

the above equations are immediately verified for the asymptotic integrals. In order to establish the horizon integrals, we have to convince ourselves that the additional terms do not contribute, i.e., that

$$\int_{\mathcal{H}} \frac{k \wedge \omega}{V} = 0, \quad \int_{\mathcal{H}} \frac{k \wedge B}{V} = 0, \quad \int_{\mathcal{H}} \frac{k \wedge E}{V} = 0. \quad (8.19)$$

We have already argued that the first integral vanishes (see eq. (8.6)). In order to verify the remaining formulae, we note that E is orthogonal to k and hence

$$(\frac{k \wedge E}{V} | \frac{k \wedge E}{V}) = \frac{1}{V^2} (k|k)(E|E) = \sigma_E^2 \quad \text{on } H[k] \quad ,$$

where we have used eq. (8.12) in the last step. (A similar expression holds for B.) Hence, $V^{-2}(k \wedge E | k \wedge E)$ remains finite on the horizon. Applying the same argument as used in the proof of theorem 8.2 yields the desired result. $\qquad \square$

We are now in a position to apply Stokes' theorem (2.8) with $\partial\Sigma = S_\infty^2 - \mathcal{H}$. Using eq. (1.23), Stokes' theorem for arbitrary stationary 1–forms α and functions $\mathcal{F}$, $L_k\alpha = L_k\mathcal{F} = 0$, implies

$$\int_{\partial\Sigma} \mathcal{F} * (k \wedge \alpha) = \int_\Sigma [(\alpha|d\mathcal{F}) - \mathcal{F} d^\dagger \alpha] i_k \eta. \quad (8.20)$$

Choosing for α the 1–forms $d\phi/V$, $d\psi/V$, dV/V and ω/V^2 and taking advantage of the basic equations (8.9)-(8.11) for the co-derivatives, and the expressions (8.16)-(8.18) for the boundary terms, we obtain the following four integral identities

$$4\pi(a_\infty Q - a_H Q_H) = -\int_\Sigma \left[\frac{(E|da)}{V} + 2a \frac{(B|\omega)}{V^2} \right] i_k \eta, \quad (8.21)$$

$$4\pi(b_\infty P - b_H P_H) = -\int_\Sigma \left[\frac{(B|db)}{V} - 2b \frac{(E|\omega)}{V^2} \right] i_k \eta, \quad (8.22)$$

$$8\pi(f_\infty M - f_H M_H) =$$

$$\int_\Sigma \left[\frac{(dV|df)}{V} + 2f \left(\frac{(E|E) + (B|B)}{V} - 2\frac{(\omega|\omega)}{V^2} \right) \right] i_k \eta, \quad (8.23)$$

$$0 = \int_\Sigma \frac{(dg|\omega)}{V^2} i_k \eta. \quad (8.24)$$

These formulae hold for arbitrary functions a, b, f and g which are algebraic expressions in the potentials ϕ, ψ, V and U. Since the latter have vanishing Lie derivatives with respect to k and assume constant values on $\mathcal{H}$, the same also holds for the former: $\mathcal{F}_H \equiv \mathcal{F}(\phi_H, \psi_H, V_H = 0, U_H)$, where $\mathcal{F} = a$, bf or g. Note that eq. (8.24) is obtained from the fact that the integral is equal to $\int_{\partial\Sigma} g * (k \wedge \frac{\omega}{V^2})$ which vanishes since g is constant on both parts of the boundary and $*(k \wedge \frac{\omega}{V^2})$ is an exact 2–form (see eq. (2.12)).

Let us now find the most general algebraic expressions a, b, f and g for which the equation

$$2f_\infty M - 2f_H M_H = a_\infty Q - a_H Q_H + b_\infty P - b_H P_H \qquad (8.25)$$

is satisfied. Considering the r.h.s. of the formulae (8.21)-(8.24) and using $d\mathcal{F} = \mathcal{F}_{,\phi} E + \mathcal{F}_{,\psi} B + \mathcal{F}_{,V} dV + \mathcal{F}_{,U} dU = (\mathcal{F}_{,\phi} - \psi\mathcal{F}_{,U})E + (\mathcal{F}_{,\psi} + \phi\mathcal{F}_{,U})B + \mathcal{F}_{,V} dV + \mathcal{F}_{,U} \omega$ for $\mathcal{F} = a$, bf and g, we obtain the following set of differential equations:

$$f_{,V} = 0, \qquad\qquad\qquad g_{,V} = -V f_{,U},$$
$$-2a = g_{,\psi} + \phi g_{,U} + V b_{,U}, \qquad 2b = g_{,\phi} - \psi g_{,U} + V a_{,U},$$
$$a_{,V} = \psi f_{,U} - f_{,\phi}, \qquad\qquad -b_{,V} = \phi f_{,U} + f_{,\psi},$$
$$2f = \psi a_{,U} - a_{,\phi}, \qquad\qquad 2f = -\phi b_{,U} - b_{,\psi},$$
$$a_{,\psi} + b_{,\phi} = \psi b_{,U} - \phi a_{,U}, \qquad g_U = 4f.$$

Solving these equations yields the following result:

Proposition 8.6 *Let a, b, f and g be the expressions*

$$a = +(\alpha\psi - \beta_1) V - (\alpha\phi + \beta_2) 2U + \tilde{a}(\phi, \psi), \qquad (8.26)$$
$$b = -(\alpha\phi + \beta_2) V - (\alpha\psi - \beta_1) 2U + \tilde{b}(\phi, \psi), \qquad (8.27)$$
$$f = \alpha U + \beta_0 + \beta_1 \phi + \beta_2 \psi, \qquad (8.28)$$
$$g = 2\alpha U^2 + [\beta_0 + \beta_1\phi + \beta_2\psi]4U - \frac{\alpha}{2}V^2 + \tilde{g}(\phi, \psi) \quad (8.29)$$

in terms of the potentials ϕ, ψ, V and U, where $\tilde{a}$, $\tilde{b}$ and $\tilde{g}$ are the following polynomials in ϕ and ψ:

$$\tilde{a} = -\alpha\psi(\phi^2 + \psi^2) + \beta_1(3\psi^2 - \phi^2) - 4\beta_2\phi\psi - 2\beta_0\phi + \gamma\psi + \delta_1,$$

$$\tilde{b} = +\alpha\phi(\phi^2 + \psi^2) + \beta_2(3\phi^2 - \psi^2) - 4\beta_1\phi\psi - 2\beta_0\psi - \gamma\phi - \delta_2,$$

$$\tilde{g} = (\phi^2 + \psi^2)[(\alpha/2)(\phi^2 + \psi^2) + 2\beta_2\phi - 2\beta_1\psi - \gamma] - 2(\delta_1\psi + \delta_2\phi).$$

Then the identity (8.25) holds for arbitrary constants α, β_0, β_1, β_2, δ_1, δ_2 and γ.

Proof The verification of this proposition is straightforward: Since the system of differential equations is linear, we can set six of the above constants equal to zero and the remaining one equal to one. For $\alpha = 1$ this yields, for instance, $a = \psi(V - \phi^2 - \psi^2) - 2U\phi$, $b = -\phi(V - \phi^2 - \psi^2) - 2U\psi$, $f = U$, $g = 2U^2 - V^2/2 + (\phi^2 + \psi^2)^2/2$, which is easily seen to solve all differential equations. $\qquad\square$

Let us now evaluate the relation (8.25) for different choices of the above constants. In the following we adopt the *gauge*

$$\phi_\infty = 0, \quad \psi_\infty = 0, \quad U_\infty = 0. \tag{8.30}$$

Setting six of the constants α, β_0, β_1, β_2, δ_1, δ_2, γ equal to zero and the remaining one equal to one, eq. (8.25) yields the following relations:

(i) $\delta_1 = 1$ or $\delta_2 = 1 \Longrightarrow a = 1$, $b = 0$, $f = 0$, $g = -2\psi$ or $a = 0$, $b = -1$, $f = 0$, $g = -2\phi$. Inserting this in eq. (8.25) gives

$$Q = Q_H, \quad P = P_H. \tag{8.31}$$

(Note that $Q = Q_H$ is a trivial consequence of Maxwell's equations in the static case. However, if neither ω nor B are assumed to vanish one has to take eq. (8.24) and $B = d\psi$ into account in order to draw this conclusion.)

(ii) $\gamma = 1 \Longrightarrow a = \psi$, $b = -\phi$, $f = 0$. This implies the relation

$$Q\,\psi_H = P\,\phi_H \tag{8.32}$$

between the values of the potentials on the horizon.

(iii) $\beta_0 = 1 \Longrightarrow a = -2\phi$, $b = -2\psi$, $f = 1$. Together with the gauge condition (8.30), eq. (8.25) now yields

$$M = M_H + \phi_H Q + \psi_H P, \tag{8.33}$$

which is the generalized Smarr formula for nonrotating electrovac black holes ($M_H = (4\pi)^{-1}\kappa A$). (Recall that we already established this relation by different means in section 7.2; see eq. (7.46).)

(iv) $\alpha = 1 \Longrightarrow a = \psi(V - \phi^2 - \psi^2) - 2U\psi$, $b = -\phi(V - \phi^2 - \psi^2) - 2U\psi$, $f = U$. Using the values of these functions on the horizon and S_∞^2, eq. (8.25) becomes $(M_H + Q\phi_H + P\psi_H)U_H = 0$, that is, $MU_H = 0$ and hence

$$U_H = 0. \tag{8.34}$$

As we shall see in the next section, the fact that $U_H = U_\infty = 0$ is of fundamental importance in order to show that, in a static domain of outer communications, the electric and magnetic potentials depend only on V.

(v) $\beta_1 = 1$ or $\beta_2 = 1 \implies a = -V + 3\psi^2 - \phi^2$, $b = 2U - 4\phi\psi$, $f = \phi$, or $a = -2U - 4\phi\psi$, $b = -V + 3\phi^2 - \psi^2$, $f = \psi$. These solutions, together with $U_H = 0$, provide the following two relations

$$Q\,(1 + 3\psi_H^2 - \phi_H^2) - P\,4\phi_H\psi_H = 2\,M_H\,\phi_H\,,$$
$$P\,(1 + 3\phi_H^2 - \psi_H^2) - Q\,4\phi_H\psi_H = 2\,M_H\,\psi_H\,. \quad (8.35)$$

Using $Q\psi_H = P\phi_H$ enables one to compute the values of the electric and the magnetic potentials on the horizon:

$$\phi_H = Q\,\frac{\sqrt{M_H^2 + Q^2 + P^2} - M_H}{Q^2 + P^2}\,, \qquad \psi_H = \phi_H\,\frac{P}{Q}\,. \quad (8.36)$$

Finally, taking advantage of these expressions in the Smarr formula (8.33), yields the expression

$$M^2 = M_H^2 + Q^2 + P^2 \quad (8.37)$$

for the total mass in terms of the charges and the horizon quantity $M_H = (4\pi)^{-1}\kappa\mathcal{A}$.

In summary, we have established the following result:

Corollary 8.7 *Consider a stationary electrovac black hole spacetime with asymptotically timelike Killing field k and nonrotating, connected horizon $H[k]$. Then the total mass and charges are subject to the relation*

$$M^2 - (\frac{\kappa}{4\pi}\mathcal{A})^2 = Q^2 + P^2\,, \quad (8.38)$$

where κ and $\mathcal{A}$ are the surface gravity and the area of $H[k]$, respectively. On $H[k]$, the combined twist potential U (defined by $dU = \omega + \phi d\psi - \psi d\phi$) and the electric and magnetic potentials assume the values

$$U_H = 0\,, \qquad \phi_H = \frac{Q}{M + M_H}\,, \qquad \psi_H = \frac{P}{M + M_H}\,, \quad (8.39)$$

where $M_H = (4\pi)^{-1}\kappa\mathcal{A}$, and the gauge is chosen such that ϕ, ψ and U vanish asymptotically.

The reasoning presented above can easily be generalized to the case where the horizon is not connected. In fact, one only has to replace eq. (8.25) by

$$2f_\infty M - \sum 2f_i M_i = a_\infty Q - \sum a_i Q_i + b_\infty P - \sum b_i P_i\,, \quad (8.40)$$

where the sum runs over the connected components of the horizon and $Q = \sum Q_i$, $P = \sum P_i$. Taking again advantage of the solutions (i)-(iv), one obtains the relations between the horizon values of the potentials and the charges. However, unless the horizon consists of exactly one component, or the values of the potentials are *required* to be equal on each component, it is no longer possible to solve for the potentials and derive a relation similar to eq. (8.38). Nevertheless, the method is likely to provide a useful tool for the uniqueness proof of the Papapetrou–Majumdar multi–black hole solution (Heusler 1995c).

8.4 Static identities and further applications

In this section we introduce two new potentials Φ and Ψ. In terms of ϕ, ψ, Φ and Ψ, the Einstein–Maxwell equations (8.9)-(8.11) assume a particularly symmetric form. Moreover, the new potentials vanish on both parts of the boundary, S^2_∞ and $\mathcal{H}$. In the static case, this enables one to conclude that the electric (and the magnetic) potential depends only on the gravitational potential V. As already mentioned, this result, first derived by Israel (1968) for the purely electric case, is of fundamental importance for the uniqueness proof of the Reissner–Nordström metric. We conclude this section with some comments on the staticity theorem for nonrotating electrovac black holes. As we argued in the second section, the latter is easy to prove if one imposes the condition $E \wedge B = 0$. However, dropping this requirement, all attempts to generalize the staticity theorem along the lines indicated by Carter (1987) and further developed in the present chapter have failed until now. We conclude this chapter by recalling Carter's argument, which works only if the inequality $(2\phi)^2 < V$ is imposed.

Let us now introduce the new potentials Φ and Ψ and establish the following proposition:

Proposition 8.8 *Let* $\hat{Q} = \frac{QM}{M^2 - M_H^2}$, $\hat{P} = \frac{PM}{M^2 - M_H^2}$ *and let* Φ *and* Ψ *be defined as*

$$\Phi = U + \hat{P}\phi - \hat{Q}\psi, \qquad (8.41)$$

$$\Psi = \frac{1}{2}\left(V - 1 - \phi^2 - \psi^2\right) + \hat{Q}\phi + \hat{P}\psi. \qquad (8.42)$$

Then the Einstein–Maxwell equations (8.9)-(8.11) become

$$d^\dagger \left(\frac{d\phi}{V} \right) = -2 \frac{(\omega | d\psi)}{V^2}, \qquad d^\dagger \left(\frac{d\psi}{V} \right) = 2 \frac{(\omega | d\phi)}{V^2}, \qquad (8.43)$$

$$d^\dagger \left(\frac{d\Phi}{V} \right) = -2 \frac{(\omega | d\Psi)}{V^2}, \qquad d^\dagger \left(\frac{d\Psi}{V} \right) = 2 \frac{(\omega | d\Phi)}{V^2}. \qquad (8.44)$$

In addition, Φ and Ψ vanish at infinity and at the horizon,

$$\Phi_\infty = \Phi_H = 0, \qquad \Psi_\infty = \Psi_H = 0. \qquad (8.45)$$

Proof Using the derivatives

$$d\Phi = \omega - (\hat{Q} - \phi) B + (\hat{P} - \psi) E, \qquad (8.46)$$

$$d\Psi = \frac{dV}{2} + (\hat{Q} - \phi) E + (\hat{P} - \psi) B, \qquad (8.47)$$

and the basic equations (8.9)-(8.11), it is a simple task to verify eqs. (8.44): For instance, the co–derivative of $d\Phi/V$ becomes

$$d^\dagger \left(\frac{d\Phi}{V} \right) = d^\dagger \left(V \frac{\omega}{V^2} \right) + d^\dagger \left[(\phi - \hat{Q}) \frac{B}{V} \right] - d^\dagger \left[(\psi - \hat{P}) \frac{E}{V} \right]$$

$$= -\frac{(\omega | dV)}{V^2} - \frac{(E|B)}{V} + 2(\phi - \hat{Q}) \frac{(E|\omega)}{V^2} + \frac{(B|E)}{V} + 2(\psi - \hat{P}) \frac{(B|\omega)}{V^2}$$

$$= -2 \frac{(\omega \,|\, [dV/2 + (\hat{Q} - \phi)E + (\hat{P} - \psi)B])}{V^2} = -2 \frac{(\omega | d\Psi)}{V^2}.$$

In a similar way one establishes the second relation in eq. (8.44). As a consequence of the gauge conditions $U_\infty = \phi_\infty = \psi_\infty = 0$ and the fact that $V_\infty = 1$, both Φ and Ψ vanish at S^2_∞. In addition, Φ vanishes on the horizon since $U_H = 0$ and $P\phi_H = Q\psi_H$ (see eqs. (8.32) and (8.34)). It remains to be verified that $\Psi_H = 0$ as well. This is, however, immediately seen from $V_H = 0$ and eqs. (8.37), (8.39).   □

As a consequence of the above relations we now obtain the well-known result that the electric and magnetic potentials depend only on V, provided that the domain is static.

Corollary 8.9 *Consider a strictly static domain of outer communications and let ϕ and ψ denote the electric and the magnetic* *potential, respectively. Then*

$$\frac{\phi}{Q} = \frac{\psi}{P} = \frac{1}{\hat{M}} \left[1 - \sqrt{1 + \frac{\hat{M}}{M}(V - 1)} \right], \qquad (8.48)$$

where $V = -(k|k)$, $\hat{M} = \frac{M^2 - M_H^2}{M} = \frac{Q^2 + P^2}{M}$ *and* $M_H = (4\pi)^{-1}\kappa \mathcal{A}$.

Proof For $\omega = 0$ eqs. (8.44) reduce to $d^\dagger(d\Phi/V) = 0$ and $d^\dagger(d\Psi/V) = 0$. Hence, using Stokes' theorem (8.20) with $f = \Psi$ and $\alpha = d\Psi/V$, we find

$$\int_{\partial\Sigma} \Psi * (k \wedge \frac{d\Psi}{V}) = \int_\Sigma \frac{(d\Psi|d\Psi)}{V} i_k \eta. \qquad (8.49)$$

It is easy to verify that $V^{-1}(k \wedge d\Psi)$ remains finite at S^2_∞ and $\mathcal{H}$. Using now the result that Ψ vanishes on the horizon and at infinity, and the fact that the same arguments apply to Φ, we have

$$\int_\Sigma \frac{(d\Phi|d\Phi)}{V} i_k \eta = \int_\Sigma \frac{(d\Psi|d\Psi)}{V} i_k \eta = 0. \qquad (8.50)$$

Since $d\Phi$ and $d\Psi$ are orthogonal to k (recall that $0 = L_k\Phi = (k|d\Phi)$), and since k is nowhere spacelike, we are able to conclude that $d\Phi$ and $d\Psi$ vanish. Hence, both Φ and Ψ are constant, i.e.,

$$\Phi = 0, \quad \Psi = 0. \qquad (8.51)$$

The assertion is now established by solving eqs. (8.41) and (8.42) (with $\Phi = \Psi = \omega = 0$) for ϕ and ψ, making also use of $Q^2 + P^2 = M\hat{M}$. $\qquad \square$

The demonstration that the electric potential is constant on the surfaces of constant V represents one of the fundamental steps in the original uniqueness proof for the Reissner–Nordström metric (Israel 1968, Müller zum Hagen *et al.* 1974). Having established this fact, it remains to prove that the surfaces of constant V have spherical symmetry. Instead of using divergence identities, the idea behind the new attempts to the static uniqueness theorem is to establish conformal flatness of the 3–geometry. From this, the desired results are then obtained as a consequence of the positive energy theorem (see below).

Let us eventually try to generalize the above corollary to the nonstatic case. First of all, we note that the r.h.s. of eqs. (8.44) can be written as co–derivatives as well, since $d^\dagger(\omega/V^2) = 0$.

This yields the relations $d^\dagger(d\Phi/V) = 2d^\dagger(\Psi\omega/V^2)$, $d^\dagger(d\Psi/V) = -2d^\dagger(\Phi\omega/V^2)$, and thus

$$d^\dagger \left[\Phi\frac{d\Phi}{V} - \Psi\frac{d\Psi}{V} - 4\Phi\Psi\frac{\omega}{V^2} \right] = \frac{(d\Psi|d\Psi)}{V^2} - \frac{(d\Phi|d\Phi)}{V^2}.$$

Using Stokes' theorem and the fact that the boundary integrals vanish, we obtain

$$\int_\Sigma \frac{(d\Psi|d\Psi)}{V^2} i_k\eta - \int_\Sigma \frac{(d\Phi|d\Phi)}{V^2} i_k\eta = 0. \qquad (8.52)$$

In a similar way one also finds $\int_\Sigma V^{-2}(d\Phi|d\Psi) = 0$ and hence

$$\int_\Sigma \frac{(d\mathcal{Z}|d\mathcal{Z})}{V^2} i_k\eta = 0, \quad \text{where} \quad \mathcal{Z} = \Psi + i\,\Phi. \qquad (8.53)$$

If the relative signs in front of the quadratic terms in eq. (8.52) were different, we would immediately be able to conclude that $\Phi = 0$ and $\Psi = 0$, as in the static case. This would also imply $\omega = 0$, i.e., it would prove the electrovac staticity theorem.

It might be instructive to mention that the Ernst potentials E and Λ defined in section 5.3 (based on the Killing field k, i.e., $N = -V$) also obey the same co–derivative equation as the complex potential $\mathcal{Z}$ introduced above:

$$d^\dagger \left(\frac{d\cdot}{V}\right) = -2i\frac{(\omega|d\cdot)}{V^2}, \quad \text{for } \mathcal{Z}, \text{ E and } \Lambda. \qquad (8.54)$$

This is seen from eq. (5.41) with $dE - dV + 2\bar{\Lambda}d\Lambda = 2i\omega$ (see also eq. (5.39)). However, the potential $\mathcal{Z}$ is more suited to establish the functional relationship (8.48) between ϕ and V since, in contrast to the Ernst potential E, it vanishes at infinity *and* on the horizon. Clearly $\mathcal{Z}$ can be expressed in terms of the potentials E and Λ, since $d\mathcal{Z} = \hat{M}^{-1} d\,[E/2 - \Lambda(Q + iP)]$.

We conclude this section by briefly reviewing Carter's (1987) staticity argument: Consider the choice $g = U + 3\phi\psi$ in eq. (8.24) and $b = \psi$ in eq. (8.22). Adding the resulting expressions gives

$$-4\pi\,\psi_H P = \int_\Sigma \frac{([B + 2\phi\omega/V]\,|\,[B + 2\phi\omega/V])}{V} i_k\eta$$
$$+ \int_\Sigma (1 - 4\frac{\phi^2}{V})\frac{(\omega|\omega)}{V^2} i_k\eta, \qquad (8.55)$$

which shows that both ω and B must vanish if the magnetic charge is zero and if ϕ and V are subject to the inequality $V > (2\phi)^2$.

However, this additional assumption has no physical justification, as is already obvious from the fact that it is violated for the *static* Reissner–Nordström solution with $4M/5 < Q < M$. Moreover, even for nonvanishing magnetic charge P, it should be possible to establish that in a strictly stationary domain of outer communications E and B are parallel, implying $\omega = 0$.

As mentioned earlier, the staticity theorem for nonrotating electrovac black holes has been established by means of the ADM formalism for a foliation by maximal slices. We feel, however, that it should be possible to prove the theorem also within the reasoning presented in this chapter, at least under the assumption of *strict* stationarity.

9

Uniqueness theorems for nonrotating holes

In this chapter we present the arguments which establish that the Schwarzschild metric describes the only static, asymptotically flat vacuum spacetime with regular (not necessarily connected) event horizon (Israel 1967, Müller zum Hagen *et al.* 1973, Robinson 1977, Bunting and Masood-ul-Alam 1987). We then discuss the generalization of this result to the situation with electric fields; that is, we demonstrate the uniqueness of the 2–parameter Reissner–Nordström solution amongst all asymptotically flat, static electrovac black hole configurations with nondegenerate horizon (Israel 1968, Müller zum Hagen *et al.* 1974, Simon 1985, Ruback 1988, Masood-ul-Alam 1992). Taking magnetic fields into account as well, we finally establish the uniqueness of the 3–parameter Reissner–Nordström metric. We conclude this chapter with a brief discussion of the Papapetrou–Majumdar metric, representing a static configuration with $M = |Q|$ and an arbitrary number of extreme black holes (Papapetrou 1945, Majumdar 1947). This metric is *not* covered by the static uniqueness theorems, since the latter apply exclusively to electrovac solutions which are subject to the inequality $M > |Q|$.

Throughout this chapter the domain of outer communications is assumed to be static. In the vacuum or the electrovac case staticity is, as we have argued in the previous chapter, a consequence of the symmetry conditions for the matter fields.

Our main objective in this chapter concerns the "modern" approach to the static uniqueness theorem, which is based on conformal transformations and the positive energy theorem. We shall, however, start this chapter with some comments on the traditional line of reasoning, which is due to Israel, Müller zum Hagen, Robinson and others.

9.1 The Israel theorem

In 1967, Israel presented the first proof for the uniqueness of the Schwarzschild metric amongst all static vacuum black hole configurations (Israel 1967). His strategy consisted in constructing two integral identities, which he used to investigate the geometric properties of the 2–surfaces $V = -(k|k) =$ constant. Assuming that the latter are regular and homeomorphic to 2–spheres, Israel was able to conclude that the solution is spherically symmetric and therefore had to be the Schwarzschild metric.

In a technically involved paper, Müller zum Hagen *et al.* (1973) succeeded in removing the regularity condition for the surfaces of constant V. In addition, they were able to justify certain topological assumptions used in Israel's paper.

A further simplification of the uniqueness proof was achieved by Robinson in 1977. Deriving a manifestly *regular* divergence formula for the square of the tensor R_{abc} (see eq. (2.48)), Robinson was able to conclude that R_{abc} vanishes. Since R_{abc} is the 3–dimensional analogue of the Weyl tensor, this implies that the hypersurfaces orthogonal to k are conformally flat (see theorem 2.12). In the vacuum case, an identity for $R_{abc}R^{abc}$ then establishes spherical symmetry.

Turning to the uniqueness proof with electromagnetic fields, the first step was again performed by Israel (1968), who generalized the results previously obtained for the vacuum case. Applying Stokes' theorem, he first derived a relation between the electric potential ϕ and the norm V of the Killing field (see corollary 8.9). Subsequently, he established spherical symmetry by similar means as in the vacuum case.

Later, Müller zum Hagen *et al.* (1974) were again able to weaken Israel's assumptions concerning the topology and the regularity of the 2–surfaces with constant V. As in the vacuum case, they used a suitably constructed divergence formula to show that R_{abc} vanishes. Again, Einstein's equations, together with an identity for $R_{abc}R^{abc}$, imply spherical symmetry.

Simon (1985) achieved a further simplification by using the $SO(2,1)$ symmetry of the static Einstein–Maxwell equations. Performing an $SO(1,1)$ transformation and using the expression for ϕ in terms of V reduces the problem to the vacuum case. The uniqueness result for the Reissner–Nordström solution is then obtained as a consequence of the corresponding vacuum theorem after restoring the $SO(1,1)$ symmetry (see also Simon 1992).

The main idea of the new approach to the static uniqueness theorems is to establish conformal flatness of the spacelike hypersurface (Σ, g). However, in contrast to the "traditional" method, the strategy is not to conclude this from ingeniously constructed divergence formulae. Instead, one looks for conformal transformations on (Σ, g), such that the transformed 3–dimensional Riemannian manifold has *non–negative* scalar curvature and *vanishing* mass (Bunting and Masood–ul–Alam 1987, Masood–ul–Alam 1993). Taking advantage of the positive mass theorem (Schoen and Yau 1979, 1981, Witten 1981; see also Horowitz and Perry 1982, Taubes and Parker 1982, Tod 1983), enables one then to conclude that (Σ, g) is conformally flat. A further advantage of this strategy is that no assumptions concerning the connectedness of the horizon enter the proof. This enables one to exclude multi black hole solutions with nondegenerate event horizon.

Before we present this reasoning, we give an outline of Israel's arguments (Israel 1967), which are based on the general formulae derived in section 2.6. The key idea is to integrate two suitably chosen combinations of Einstein's vacuum equations over the spacelike hypersurface Σ and to compare the result with the Komar formula (that is, with the integrated Poisson equation). Using the results and notations of section 2.6, the relevant combinations of Einstein's equations turn out to be $G_{\hat{0}\hat{0}} + G_{\hat{1}\hat{1}}$ and $G_{\hat{0}\hat{0}} + 3G_{\hat{1}\hat{1}}$. (In order to avoid ambiguities, we equip tensor indices referring to the orthonormal tetrad fields $\theta^0 = S dt$ and $\theta^1 = \rho dS$ with a hat. Throughout this chapter we use the notation $S^2 \equiv V = -(k|k)$ and assume that the surfaces of constant S are regular.) Using eqs. (2.52), (2.53) and (2.62)-(2.63), we obtain

$$
\begin{aligned}
G_{\hat{0}\hat{0}} + G_{\hat{1}\hat{1}} &= S^{-2}G_{tt} + \rho^{-2}G_{SS} = \frac{1}{2}R^{(g)} + \rho^{-2}G_{SS}^{(g)} + \frac{K}{\rho S} \\
&= \frac{1}{\rho}\left[\frac{K}{S} - \partial_S K - \frac{\rho}{2}K^2\right] - \frac{2}{\sqrt{\rho}}\tilde{\Delta}\sqrt{\rho} \\
&\quad - \left[\frac{(\tilde{\nabla}\rho | \tilde{\nabla}\rho)}{2\rho^2} + \overset{\circ}{K}_{ab}\overset{\circ}{K}{}^{ab}\right],
\end{aligned} \tag{9.1}
$$

where $\overset{\circ}{K}_{ab}$ denotes the trace–free part of the extrinsic curvature of the 2–surfaces $S = $ constant in Σ, $\overset{\circ}{K}_{ab} = K_{ab} - \frac{1}{2}\tilde{g}_{ab}K$. (We recall that $\boldsymbol{g}$ denotes the 3–dimensional metric on Σ and that $\tilde{g}$ is the induced metric on the 2–surfaces $S = $ constant.) In a similar way one finds

$$G_{\hat{0}\hat{0}} + 3\,G_{\hat{1}\hat{1}} = \frac{1}{\rho}\left[3\frac{K}{S} - \partial_S K\right] - \tilde{R} - \tilde{\Delta}\ln\rho$$
$$- \left[\frac{(\tilde{\nabla}\rho|\tilde{\nabla}\rho)}{\rho^2} + 2\,\overset{\circ}{K}_{ab}\overset{\circ}{K}{}^{ab}\right]. \qquad (9.2)$$

In order to complete the list of relevant equations, we recall the Poisson equation (2.67),

$$\partial_S\,\rho = \rho^2\,(K - \frac{\rho}{S}R_{tt}), \qquad (9.3)$$

and the equation for $\partial_S\sqrt{\tilde{g}}$, which is a consequence of the definition (2.57) of the extrinsic curvature:

$$\partial_S\sqrt{\tilde{g}} = \sqrt{\tilde{g}}\,K\,\rho. \qquad (9.4)$$

Israel's theorem is now obtained from the vacuum equations $G_{\hat{0}\hat{0}} = 0$, $G_{\hat{1}\hat{1}} = 0$ and $R_{tt} = 0$ as follows: Using the consequence $\partial_S(\sqrt{\tilde{g}}/\sqrt{\rho}) = \sqrt{\tilde{g}}\sqrt{\rho}K/2$ of eqs. (9.3) and (9.4), one finds for the first term in brackets in eq. (9.1)

$$\left[\frac{K}{S} - \partial_S K - \frac{\rho}{2}K^2\right] = -S\frac{\sqrt{\rho}}{\sqrt{\tilde{g}}}\,\partial_S\left(\frac{\sqrt{\tilde{g}}}{\sqrt{\rho}}\frac{K}{S}\right),$$

which yields the inequality

$$\partial_S\left(\frac{\sqrt{\tilde{g}}}{\sqrt{\rho}}\frac{K}{S}\right) \leq -2\frac{\sqrt{\tilde{g}}}{S}\tilde{\Delta}\sqrt{\rho}, \qquad (9.5)$$

where equality holds if and only if both $\overset{\circ}{K}_{ab}$ and $\tilde{\nabla}\rho$ vanish. In a completely analogous manner one extracts the inequality

$$\partial_S\left(\frac{\sqrt{\tilde{g}}}{\rho}[KS + \frac{4}{\rho}]\right) \leq -S\sqrt{\tilde{g}}\,(\tilde{\Delta}\ln\rho + \tilde{R}) \qquad (9.6)$$

from the second combination (9.2) where, as before, equality holds iff $\overset{\circ}{K}_{ab} = \tilde{\nabla}\rho = 0$.

Before we are able to integrate the above inequalities over Σ, we must investigate the asymptotic properties and the behavior of the various quantities near the horizon. In order to do so, we first note that on the horizon ρ^{-1} coincides with the surface gravity. One way to see this is to evaluate the expression (6.16) for κ in the metric given by eqs. (2.38) and (2.56). With $k = -S^2 dt$ we have $(dk|dk) = 4S^2(dS|dS)(dt|dt)\,|_{H[k]} = -4/\rho_H$, which establishes the desired result, $\kappa = 1/\rho_H$. In particular, this implies that ρ_H is constant with $0 < \rho_H < \infty$, provided that we deal with a regular Killing horizon. In addition, eq. (2.72) implies that K_{ab} and K vanish on the horizon if $R_{\alpha\beta\gamma\delta}R^{\alpha\beta\gamma\delta}$ is required to remain finite. Now using the $G_{\hat{1}\hat{1}}$ equation (see eqs. (2.53), (2.62) and (2.63)) we have

$$K_{ab} = 0, \qquad \frac{K}{S} = \frac{1}{2}\rho\tilde{R} \quad \text{on } H[k]. \tag{9.7}$$

Turning to the asymptotic behavior, one assumes the existence of an asymptotic coordinate system $\{x^\alpha\}$ in which the metric has the usual asymptotically flat form,

$$S^2 = 1 - \frac{2M}{r} + \mathcal{O}(r^{-2}),$$

$$g_{ij} = (1 + \mathcal{O}(r^{-1}))\delta_{ij} + \mathcal{O}(r^{-2}), \tag{9.8}$$

where $r^2 = \delta_{\mu\nu}x^\mu x^\nu$. Using the fact that $\rho^{-2} = (dS|dS)$, we find in leading order

$$\rho^{-1} \to \frac{M}{r^2}, \qquad K \to \frac{2}{r}, \qquad \text{as } r \to \infty. \tag{9.9}$$

We are now ready to integrate the above inequalities over Σ, which extends from its inner boundary $\mathcal{H}$ $(S = 0)$ to spacelike infinity S^2_∞ $(S = 1)$. In order to compute the right hand sides, we note that the Gauss–Bonnet theorem yields for the topological 2–surfaces given by $S = $ constant,

$$\int_\Sigma S\tilde{R}\,dS \wedge \tilde{\eta} = \int_0^1 S\,dS \int_S \tilde{R}\tilde{\eta} = \frac{1}{2}8\pi. \tag{9.10}$$

In addition, Stokes' theorem implies

$$\int_S \tilde{\Delta}\sqrt{\rho}\,\tilde{\eta} = \int_S d\tilde{*}d\sqrt{\rho} = 0 \tag{9.11}$$

and, in a similar way, $\int_S \tilde{\Delta}\ln\rho\,\tilde{\eta} = 0$. Thus, the inequalities (9.5) and (9.6) yield the following estimates:

$$\left[\int_S \frac{K}{\sqrt{\rho}S}\tilde{\eta}\right]_0^1 \leq 0, \qquad \left[\int_S \frac{KS + 4\rho^{-1}}{\rho}\tilde{\eta}\right]_0^1 \leq -4\pi. \tag{9.12}$$

Now using the properties (9.7) and (9.9), and again the Gauss–Bonnet theorem, one obtains $8\pi\sqrt{M} - 4\pi\sqrt{\rho_H}$ for the first integrand, and $-4\mathcal{A}\rho_H^{-2}$ for the second one. (As usual, $\mathcal{A}$ denotes the area of $\mathcal{H} = \Sigma \cap H[k]$.) This eventually gives the estimates

$$M \le \frac{1}{4}\rho_H, \quad \frac{1}{4\pi}\mathcal{A}\rho_H^{-1} \ge \frac{1}{4}\rho_H, \tag{9.13}$$

where we recall that equality holds if and only if $\overset{\circ}{K}_{ab} = 0$ and $\tilde{\nabla}\rho = 0$. In order to complete the argument, we use the Komar expression for the total mass. Since, in a vacuum spacetime, the Komar integrals over $\mathcal{H}$ and S_∞^2 have the same value ($M = M_H = \frac{\kappa}{4\pi}\mathcal{A}$), we conclude from the above estimates and $\kappa^{-1} = \rho_H$ that

$$\frac{1}{4}\rho_H \ge M = \frac{\rho_H^{-1}}{4\pi}\mathcal{A} \ge \frac{1}{4}\rho_H, \tag{9.14}$$

which yields a contradiction unless equality holds everywhere. Hence, the inequalities (9.5) and (9.6) become equalities as well, implying that

$$K_{ab} - \frac{1}{2}\tilde{g}_{ab}K = 0, \quad \tilde{\nabla}\rho = 0. \tag{9.15}$$

From this, one finally concludes that the 2–surfaces $S = $ constant are spherically symmetric (see also Künzle 1971): Using eq. (9.15) in eqs. (9.1) and (9.3) with $G_{\hat{0}\hat{0}} = G_{\hat{1}\hat{1}} = R_{tt} = 0$, shows that ρ and K depend only on S. Equation (9.2) then implies that $\tilde{R}$ is constant on the surfaces of constant S. Explicitly one finds

$$\rho = \frac{4c}{(1 - S^2)^2}, \quad K = \frac{S}{c}(1 - S^2), \quad \tilde{R} = \frac{(1 - S^2)^2}{2c^2}, \tag{9.16}$$

where c is a constant of integration. (It is easy to prove that this is the only solution of eqs. (9.1) and (9.3) with $\tilde{R} \ne 0$.) Hence, defining $r(S)$ by the relation $\tilde{R} = 2/r^2$, yields (with $c = M$)

$$S^2 = 1 - \frac{2M}{r}, \quad \rho^2 dS^2 = (1 - \frac{2M}{r})^{-1}dr^2, \quad \tilde{g} = r^2 d\Omega^2, \tag{9.17}$$

which is the familiar form of the Schwarzschild metric.

The main flaw of the above reasoning lies in the regularity requirement for the embedded 2–surfaces. As already mentioned, Müller zum Hagen et al. (1973) succeeded in removing the condition $(dS|dS) \ne 0$. Later on, Robinson (1977) found a different proof which was based on an ingeniously constructed divergence identity for the square of R_{kij}. In 1987, Bunting and Masood–ul–Alam presented yet another proof of the static uniqueness theorem, making essential use of the positive energy theorem, which had been established in the intervening time. The following sections, discussing the uniqueness of the Schwarzschild and the Reissner–Nordström metrics, are devoted to this method.

9.2 Uniqueness of the Schwarzschild metric

We now present the proof of Bunting and Masood–ul–Alam (1987) establishing the uniqueness of the Schwarzschild metric amongst all asymptotically flat, static vacuum solutions with regular event horizon. First, we recall that the metric of a static spacetime can be written in the form (see section 2.5)

$$^{(4)}\boldsymbol{g} = -S^2 dt^2 + \boldsymbol{g}, \tag{9.18}$$

where $S^2 = -(k|k)$ and k denotes the stationary Killing field. The general identities (2.45)-(2.47) for the Ricci tensor in terms of S and the covariant derivative $\nabla^{(g)}$ of the 3–dimensional manifold $(\Sigma, \boldsymbol{g})$, together with the vacuum Einstein equations $R_{\mu\nu} = 0$, yield the equations

$$R^{(g)}_{ij} = S^{-1} \nabla^{(g)}_j \nabla^{(g)}_i S, \tag{9.19}$$

$$R^{(g)} = 0, \quad \Delta^{(g)} S = 0. \tag{9.20}$$

The following notion of asymptotic flatness is adopted: Let $\mathcal{K} \in \Sigma$ be a compact set, such that $\Sigma - \mathcal{K}$ is diffeomorphic to $\mathbb{R}^3 - \overline{D}$ ($\overline{D}$ being the closed unit ball centered at the origin in $\mathbb{R}^3$), and such that $\boldsymbol{g} = \boldsymbol{\delta} + \mathcal{O}(|y|^{-1})$ and $S = 1 - M|y|^{-1} + \mathcal{O}(|y|^{-2})$ in $\Sigma - \mathcal{K}$ (with respect to standard coordinates $\{y^i\}$ of $\mathbb{R}^3$). Then there exists an adapted coordinate system $\{x^i\}$ such that

$$\boldsymbol{g} = (1 + \frac{2M}{r}) \boldsymbol{\delta} + \mathcal{O}(r^{-2}), \quad S = 1 - \frac{M}{r} + \mathcal{O}(r^{-2}), \tag{9.21}$$

where the first derivatives of the terms of $\mathcal{O}(r^{-2})$ are of $\mathcal{O}(r^{-3})$ and M denotes the total mass (see Simon and Beig 1983, Simon 1984, Bunting and Masood–ul–Alam 1987, Kennefick and Ó Murchadha 1995).

We next consider conformal transformations of the 3–dimensional Riemannian manifold $(\Sigma, \boldsymbol{g})$: Let

$$\hat{\boldsymbol{g}} = \Omega^2 \boldsymbol{g}. \tag{9.22}$$

Then the Ricci scalar $\hat{R}$ assigned to the conformally transformed metric $\hat{g}$ becomes (see, e.g., Wald 1984)

$$\frac{\Omega^4}{2}\hat{R} = \frac{\Omega^2}{2}R^{(g)} - 2\Omega\Delta^{(g)}\Omega + (d\Omega|d\Omega)^{(g)}. \qquad (9.23)$$

In addition, the second fundamental form L of a 2–dimensional hypersurface $^{(2)}\Sigma$ transforms according to

$$\hat{L} = \Omega L + \frac{(n|d\Omega)^{(g)}}{\Omega^2}\,^{(2)}\hat{g}. \qquad (9.24)$$

Here $^{(2)}\hat{g}$ and n denote the induced metric from $\hat{g}$ on $^{(2)}\Sigma$ and the outward unit normal of $^{(2)}\Sigma$, respectively. We are now ready to prove the following proposition:

Proposition 9.1 *Consider the vacuum equations (9.19), (9.20) with asymptotic behavior (9.21). Let $\hat{\Sigma}_+$ and $\hat{\Sigma}_-$ denote two copies of Σ with boundaries $\mathcal{H}_+$ and $\mathcal{H}_-$ and metrics $\hat{g}_+$ and $\hat{g}_-$, respectively,*

$$\hat{g}_\pm = \Omega_\pm^2 g, \quad \text{where } \Omega_\pm = \frac{1}{4}(1 \pm S)^2. \qquad (9.25)$$

Let $\hat{\Sigma} = \hat{\Sigma}_+ \cup \hat{\Sigma}_- \cup \{P\}$, where $\hat{\Sigma}_+$ and $\hat{\Sigma}_-$ are pasted together along $\mathcal{H}_+$ and $\mathcal{H}_-$, and $\{P\}$ denotes the point at infinity. Then

(i) $\Omega_\pm$ vanishes nowhere in Σ;
(ii) $\hat{g}$ and $\hat{L}$ match continuously on $\mathcal{H}_\pm$;
(iii) $\hat{g}_-$ is extensible to $\hat{\Sigma}_- \cup \{P\}$;
(iv) the Ricci curvature $\hat{R}$ vanishes identically on $\hat{\Sigma}$;
(v) $\hat{\Sigma}_+$ is asymptotically flat with vanishing mass.

Proof

(i) Clearly $\Omega_+ \geq \frac{1}{4}$ since the domain of outer communications is assumed to be strictly stationary, i.e., $S > 0$ in Σ. As for Ω_-, we first note that the general equation (6.14) (with $N = -S^2$) yields $(dS|dS) = -\kappa^2$ on the horizon. Hence, the outward unit normal of the horizon becomes $n = -\kappa^{-1}dS$. The regularity of the horizon ($\kappa \neq 0$) then implies that $\frac{\partial(S-1)}{\partial n} < 0$ on $\mathcal{H}$. On the other hand, as $r \to \infty$, we have $(S - 1) = -M/r + \mathcal{O}(r^{-2})$. Since $(S - 1)$ is harmonic with respect to g (see eq. (9.20)), we conclude that $(S-1)$ attains its maximal value of zero only at infinity, implying that $\Omega_- \neq 0$ in Σ.

(ii) Since $S = 0$ on the horizon, the metric of $\hat{\Sigma}$ is continuous on $\mathcal{H}_{\pm}$. Since, in addition, the horizon has vanishing shear and expansion, the first term in eq. (9.24) vanishes on the 2–surface $\mathcal{H}_{\pm}$. Using $d\Omega_{\pm} = \frac{1}{2}(S \pm 1)dS$ and $n = -\kappa^{-1}dS$, eq. (9.24) yields

$$\hat{L}_{\pm} = \left[\frac{S \pm 1}{2\Omega_{\pm}^2} (n|dS)^{(g)} \right]_{S=0} \times \,^{(2)}\hat{g} = \mp 8\kappa \,^{(2)}\hat{g} \qquad (9.26)$$

for the second fundamental form with respect to the conformal metric $\hat{g}_{\pm}$. Hence, the metric and the second fundamental form match continuously on $\mathcal{H}_{\pm}$. (As a matter of fact, the metric of $\hat{\Sigma}$ is $C^{1,1}$ in the vicinity of $\mathcal{H}_{\pm}$ since the second derivatives of S and g are bounded with respect to Gaussian normal coordinates.)

(iii) Using the asymptotic expansion (9.21) and the coordinates $z^i = x^i / |x|^2$, one can show that the metric

$$\hat{g}_{-}^{\epsilon} = \begin{cases} \hat{g}_{-}(z) & \text{for } 0 < |z| < \epsilon \\ (M/2)^4 \, \delta_{ij} dz^i dz^j & \text{at } P \end{cases}$$

compactifies $\hat{\Sigma}_{-}$ in a $C^{1,1}$ manner on adding $\{P\}$ (see Bunting and Masood–ul–Alam 1987 for details).

(iv) Substituting the expression (9.25) for the conformal factor into the equation for the transformed Ricci scalar (9.23), we obtain

$$\frac{\Omega_{\pm}^4}{2} \hat{R} = \frac{\Omega_{\pm}^2}{2} R^{(g)} - \Omega_{\pm} (S \pm 1) \Delta^{(g)} S = 0, \qquad (9.27)$$

where we have used the vacuum equations (9.20) in the last step. Hence, $\hat{\Sigma}$ has vanishing scalar curvature.

(v) Using the expansion of S, we find $\Omega_{+} = 1 - M/r + \mathcal{O}(r^{-2})$ as $r \to \infty$. Together with eq. (9.21) for g, this yields

$$\hat{g}_{+} = \Omega_{+}^2 g = 1 + \mathcal{O}(r^{-2}), \quad \text{as } r \to \infty, \qquad (9.28)$$

where the term of $\mathcal{O}(r^{-1})$ vanishes. Hence, $\hat{\Sigma}_{+}$ is asymptotically flat and has vanishing mass.   □

Following Bunting and Masood–ul–Alam, we now apply the following corollary of the positive mass theorem (Schoen and Yau 1979, Witten 1981):

Theorem 9.2 *Let (N, h) be an asymptotically flat (in the sense discussed above), complete, orientable 3–dimensional Riemannian manifold with non–negative scalar curvature and vanishing mass. Then (N, h) is isometric to $(\mathbb{R}^3, \delta)$.*

It is worth pointing out that this theorem cannot be applied to the 3–dimensional manifold $(\hat{\Sigma}_+, \hat{g}_+)$. However, the requirements of the above theorem are fulfilled for the Riemannian manifold $(\hat{\Sigma}, \hat{g})$ which is composed of the two copies $\hat{\Sigma}_\pm$ of Σ. As a consequence of the positive mass theorem and the preceding proposition, one now obtains the following result:

Corollary 9.3 *The spatial geometry* (Σ, g) *of the domain of outer communications of an asymptotically flat, static vacuum spacetime with regular event horizon is conformally flat. The flat metric is* $\frac{1}{16}(1 + S)^4 g$, *where* $S^2 = -(k|k)$ *and* k *denotes the hypersurface orthogonal Killing field.*

The last step in the proof of the uniqueness theorem for the Schwarzschild metric consists in the task to establish spherical symmetry from the vacuum field equations and the fact that (Σ, g) is conformally flat. We already recalled theorem 2.12, according to which the third–rank tensor R_{kij} vanishes if and only if (Σ, g) is conformally flat. Using the vacuum field equations (9.19) and (9.20), we derived the expression

$$R_{kij} = \frac{4}{S^2} S_{|[j} S_{|i]|k} + \frac{2}{S^2} g_{k[i} S_{|j]|n} S^{|n} \qquad (9.29)$$

for R_{kij} in terms of S and the covariant derivative with respect to g. In order to proceed, we now investigate the consequences of $R_{kij} = 0$ for the 2–surfaces $^{(2)}\Sigma$ with constant S. This may be done by introducing the following quantities characterizing the geometry of $^{(2)}\Sigma$. The unit normal to $^{(2)}\Sigma$ is

$$n = w^{-1/2} dS, \quad \text{where} \quad w \equiv (dS|dS)^{(g)}, \qquad (9.30)$$

and the induced metric from g becomes

$$\beta_{ij} = g_{ij} - n_i n_j. \qquad (9.31)$$

The extrinsic curvature H_{ij} and its trace–free part $\overset{\circ}{H}_{ij}$ are given by

$$H_{ij} = \beta_i{}^m \beta_j{}^k n_{k|m}, \quad \overset{\circ}{H}_{ij} = H_{ij} - \frac{1}{2} H \beta_{ij}, \qquad (9.32)$$

where $H = H_{ij}\beta^{ij}$. A straightforward, but lengthy computation yields the useful *identities* (not involving Einstein's equations)

$$\overset{\circ}{H}_{ij}\overset{\circ}{H}{}^{ij} + \frac{1}{8w^2}\beta^{ij}\,w_i w_j$$

$$= \frac{1}{w}\left[S_{ij}S^{ij} - \frac{1}{2}(\Delta^{(g)}S)^2\right] + \frac{1}{w^2}S^i S^j S_{ij}\Delta^{(g)}S - \frac{3}{2w^3}S^i S^j S^k_{\;j}S_{ki}$$

$$= \frac{S^4}{8w^2}\left[\frac{4}{S^2}S_{[j}S_{i]k} + \frac{2}{S^2}g_{k[i}S_{j]n}S^n\right]^2 - \frac{1}{2w}(\Delta^{(g)}S)^2\,, \qquad (9.33)$$

where $w_i \equiv w_{|i}$, $S_{ij} \equiv S_{|i|j}$ and $(t_{ijk})^2 \equiv t_{ijk}t^{ijk}$. Now using the vacuum expression (9.29) for R_{kij} and the fact that $\Delta^{(g)}S = 0$, we obtain the relation

$$\frac{S^4}{8w^2}R_{kij}R^{kij} = \overset{\circ}{H}_{ij}\overset{\circ}{H}{}^{ij} + \frac{1}{8w^2}\beta^{ij}\,w_i w_j\,, \qquad (9.34)$$

which is valid if the vacuum equations hold. The conformal flatness of $(\Sigma, \boldsymbol{g})$ enables us now to conclude that

$$H_{ij} - \frac{1}{2}H\,\beta_{ij} = 0\,, \qquad \beta^{ij}\,w_{;j} = 0\,, \qquad (9.35)$$

which implies spherical symmetry. Thus, Σ is diffeomorphic to $S^2 \times \mathbb{R}$ and the metric on Σ can be written as $\boldsymbol{g} = f^2 dS^2 + r^2 d\Omega^2$, with f and r depending on the radial coordinate S only. With respect to this metric, the Poisson equation (9.20) for S and the R_{SS}–equation (9.19) become, respectively ($f' = \partial f/\partial S$)

$$(r^2 f^{-1})' = 0\,, \qquad (\frac{r'}{r})' - \frac{f'r'}{fr} + (\frac{r'}{r})^2 = \frac{1}{2S}\frac{f'}{f}\,.$$

Together with the asymptotic expansion for S, we obtain the solution $f = \frac{r^2}{M}$, $r(S) = \frac{2M}{1-S^2}$ and thus, with $f^2 dS^2 = S^{-2}dr^2$,

$$^{(4)}\boldsymbol{g} = -S^2 dt^2 + S^{-2}dr^2 + r^2 d\Omega^2\,, \qquad S^2 = 1 - \frac{2M}{r}\,. \qquad (9.36)$$

This concludes the uniqueness proof for the Schwarzschild metric (Schwarzschild 1916a, 1916b). We point out that no assumptions concerning the connectedness of the horizon enter the above reasoning. Note, however, that the arguments apply only to *regular* Killing horizons, $\kappa \neq 0$. To summarize, we have established the following theorem (Bunting and Masood–ul–Alam 1987):

Theorem 9.4 *The Schwarzschild metric represents the only vacuum black hole solution with regular event horizon and static, asymptotically flat domain of outer communications.*

9.3 Uniqueness of the Reissner–Nordström metric

We shall now generalize the results obtained in the previous section to electrically charged black holes; that is, we shall prove the uniqueness of the Reissner–Nordström metric amongst all asymptotically flat, static *electrovac* spacetimes with regular event horizon. We do so by again applying the positive mass theorem in order to establish conformal flatness of (Σ, g) (Ruback 1988, Masood-ul-Alam 1992). A similar argument as was given before is then used to derive spherical symmetry from $R_{kij} = 0$ (Müller zum Hagen *et al.* 1974).

In this section we restrict ourselves to black holes with vanishing magnetic charge, $P = 0$. In the static case this implies that the configuration is purely electric. This is immediately seen from eq. (8.22), which for $\omega = 0$, $b = \psi$ and $V = S^2 = -(k|k)$ yields (with $P_H = P$, see eq. (8.31))

$$4\pi \, \psi_H P = \int_\Sigma \frac{(B|B)}{S^2} \, i_k \eta \,. \tag{9.37}$$

Since B is orthogonal to the (nowhere spacelike) field k, this implies that

$$B = 0 \,, \quad \text{if} \quad P = 0, \quad \omega = 0 \tag{9.38}$$

and if the domain is strictly static. As in the vacuum case, we assume that there exists a compact set $\mathcal{K} \in \Sigma$ such that $\Sigma - \mathcal{K}$ is diffeomorphic to $I\!R^3 - \overline{D}$, and such that g and S have the same asymptotic behavior as before. With respect to standard coordinates induced by this diffeomorphism on $\Sigma - \mathcal{K}$, one can establish the asymptotic expansion (9.21) for the metric, and the asymptotic expansion

$$\phi = \frac{Q}{r} + \mathcal{O}(r^{-2}) \tag{9.39}$$

for the electric potential. As usual, M and Q denote the total mass and the total electric charge, respectively, and the first derivatives of the terms of $\mathcal{O}(r^{-2})$ are required to be of $\mathcal{O}(r^{-3})$. Recall that the electric potential, ϕ, is defined by $d\phi = E = -i_k F$, where the gauge is chosen such that ϕ vanishes at spacelike infinity. In terms of ϕ, the stress–energy tensor becomes (with $\phi_i \equiv \nabla_i^{(g)} \phi$

and $(\cdot|\cdot) \equiv (\cdot|\cdot)^{(g)})$

$$T_{tt} = \frac{1}{8\pi}\,(d\phi|d\phi)\,, \qquad T_{ij} = \frac{S^{-2}}{8\pi}\,[g_{ij}(d\phi|d\phi) - 2\,\phi_i\phi_j]\,, \quad (9.40)$$

implying, of course, $R^\mu_{\ \mu} = 0$. Together with Einstein's equations, the basic relations (2.45)-(2.47) now become

$$\Delta^{(g)}S = S^{-1}\,(d\phi|d\phi)\,, \tag{9.41}$$

$$R^{(g)}_{ij} = S^{-1}\nabla^{(g)}_j\nabla^{(g)}_i S + S^{-2}[g_{ij}(d\phi|d\phi) - 2\phi_i\phi_j]\,, \tag{9.42}$$

$$R^{(g)} = 2\,S^{-1}\,\Delta^{(g)}S\,. \tag{9.43}$$

In addition, we have Maxwell's equation (8.9) for ϕ, which (with $V = S^2$) in the static case becomes $2S^{-1}(d\phi|dS) = \Delta\phi = \Delta^{(g)}\phi + S^{-1}(d\phi|dS)$; that is,

$$\Delta^{(g)}\phi = S^{-1}\,(d\phi|dS)\,. \tag{9.44}$$

Again, the key to the uniqueness proof lies in the fact that there exists a conformal transformation of (Σ, g) such that the transformed 3-dimensional Riemannian manifold becomes flat. More precisely, we have the following proposition, generalizing the corresponding proposition of the previous section:

Proposition 9.5 *Consider the electrovac equations (9.41)-(9.44) with asymptotic behavior (9.21), (9.39). Let $\hat{\Sigma}_+$, $\hat{\Sigma}_-$, $\hat{\Sigma}$ and $\hat{g}_\pm$ be defined as in proposition 9.1, where now*

$$\Omega_\pm = \frac{1}{4}\,[(1\pm S)^2 - \phi^2]\,. \tag{9.45}$$

Then, provided that $|Q| < M$, the assertions (i)-(iii) and (v) of proposition 9.1 remain true. In addition, (iv): the Ricci-curvature $\hat{R}$ is non-negative.

Proof

(i) Since S is not negative in Σ, it is sufficient to show that $(1 - S)^2 - \phi^2 = (S + \phi - 1)(S - \phi - 1)$ is positive. The following argument establishes that both factors are negative in the interior of Σ: First, the Poisson equations (9.41) and (9.44) for S and ϕ imply

$$\Delta^{(g)}(S\pm\phi-1) \mp S^{-1}\,(d\phi\,|\,d[S\pm\phi-1]) = 0\,. \tag{9.46}$$

Secondly, $\kappa \neq 0$ implies $\frac{\partial(S \pm \phi - 1)}{\partial n} < 0$ with respect to the outward unit normal $n = -\kappa^{-1} dS$. This is a consequence of the relation (6.14) and $d\phi = \sigma_E k$ on $H[k]$ (see eq. (8.12)). In addition, the asymptotic expansions of the two factors are

$$S \pm \phi - 1 = -\frac{M \mp Q}{r} + \mathcal{O}(r^{-2}), \qquad \text{as } r \to \infty.$$

Applying the maximum principle and using $|Q| < M$, enables one to conclude that both factors are nonpositive in Σ and vanish only at infinity. (Note that the argument does not work if $|Q| = M$, reflecting the fact that the Papapetrou–Majumdar metric is not covered by these considerations.)

(ii) Again $\Omega_+ = \Omega_-$ for $S = 0$, which shows that the metric is continuous on $\mathcal{H}_\pm$. Since the horizon is totally geodesic, the first term in eq. (9.24) vanishes. In order to evaluate the second term, we use $d\phi = \sigma_E k$, $n = -\kappa^{-1} dS$ and the fundamental relation $S dS = \kappa k$ on $\mathcal{H}_\pm$ (see eq. (6.14)). This yields

$$\hat{L}_\pm = \left[\frac{S \pm 1 - \sigma_E \kappa^{-1} S}{2 \Omega_\pm^2} (n|dS) \right]_{S=0} \times \,^{(2)}\hat{g} = \mp 8\,\kappa\,^{(2)}\hat{g}, \quad (9.47)$$

which is the same result as in the vacuum case (see eq. (9.26)).

(iii) As before, an asymptotic expansion shows that the metric $\hat{g}_-^\epsilon$ defined in the previous section compactifies $\hat{\Sigma}_-$ in a $C^{1,1}$ manner.

(iv) In order to prove that the conformally transformed metric has non–negative Ricci curvature, we use

$$d\Omega_\pm = \frac{1}{2}[(S \pm 1)\, dS - \phi\, d\phi],$$

$$\Delta^{(g)}\Omega_\pm = \frac{1}{2}[(dS|dS) - (d\phi|d\phi) + (S \pm 1)\,\Delta^{(g)}S - \phi\,\Delta^{(g)}\phi],$$

and the field equations (9.41), (9.43) and (9.44) to substitute for the Laplacians of S and ϕ. After some straightforward algebra, the relation (9.23) for the conformally transformed Ricci scalar becomes ($\Omega \equiv \Omega_\pm$)

$$\frac{\Omega^4}{2}\hat{R} = \frac{1}{S^2}\left[\Omega^2 \mp \Omega S + \frac{1}{4}S^2\phi^2\right]|d\phi|^2 + \frac{1}{4}\phi^2\,|dS|^2$$

$$+ \frac{\phi}{4S}(1 - S^2 - \phi^2)\,(d\phi|dS). \qquad (9.48)$$

Since the term in the first bracket is equal to $[(1 - S^2 - \phi^2)/4]^2$, it is easy to see that the r.h.s. of eq. (9.48) is a perfect square, establishing that $\hat{\Sigma}$ has non–negative scalar curvature:

$$\frac{\Omega^4}{2} \hat{R} = \frac{1}{(4S)^2} \left[(1 - S^2 - \phi^2) \, d\phi + 2 \, \phi S \, dS \right]^2 . \tag{9.49}$$

(v) As in the vacuum case, $\Omega_+ = 1 - M/r + \mathcal{O}(r^{-2})$ as $r \to \infty$, which, together with the asymptotic expansion for the 3–metric, g, again yields

$$\hat{g}_+ = \Omega_+^2 g = 1 + \mathcal{O}(r^{-2}), \quad \text{as } r \to \infty, \tag{9.50}$$

establishing that $\hat{\Sigma}_+$ is asymptotically flat and has vanishing mass. $\square$

The corollary (theorem 9.2) to the positive mass theorem may now be applied to the 3–dimensional Riemannian manifold (Σ, g) constructed above. This yields the result:

Corollary 9.6 *The spatial geometry (Σ, g) of the domain of outer communications of an asymptotically flat, static, electrovac space-time with regular event horizon and vanishing magnetic charge is conformally flat. The flat metric is $\frac{1}{16}((1 + S)^2 - \phi^2)^2 \, g$, where $S^2 = -(k|k)$ and k denotes the hypersurface orthogonal Killing field. In addition, the electric and gravitational potentials are subject to the relation*

$$S^2 - \phi^2 + 2\frac{M}{Q} \phi = 1. \tag{9.51}$$

The last statement in the above corollary is obtained from $\hat{R} = 0$ and eq. (9.49) after integrating $(1 - S^2 - \phi^2)\nabla\phi = -2\phi S \nabla S$ and using the asymptotic relation $S^2 = 1 - 2\frac{M}{Q}\phi + \mathcal{O}(\phi^2)$. (In section 8.4 we also derived eq. (9.51) for connected horizons by the means of divergence identities; see eq. (8.48) with $\hat{M} = Q^2/M$ $(P = 0)$ and $V = S^2$.)

In order to conclude the uniqueness proof, it remains to be demonstrated that spherical symmetry is obtained from the electrovac field equations (9.41)-(9.44) and the conformal flatness of (Σ, g), i.e., from the relation (9.51). This is achieved by the same technique as in the vacuum case. The proof requires, however, some more algebra. We start by eliminating ϕ from the electrovac field equations, using the consequence $d\phi = SdS/(\phi - m)$ of eq. (9.51) $(m \equiv M/Q)$. This yields

$$\Delta^{(g)}S = \frac{S}{S^2 - 1 + m^2}(dS|dS), \qquad (9.52)$$

$$R_{ij}^{(g)} = \frac{1}{S}S_{ij} + \frac{1}{S^2 - 1 + m^2}[g_{ij}(dS|dS) - 2\,S_i S_j], \qquad (9.53)$$

$$R^{(g)} = \frac{2}{S^2 - 1 + m^2}(dS|dS), \qquad (9.54)$$

where, as before, $S_{ij} \equiv S_{|i|j} \equiv \nabla_j^{(g)}\nabla_i^{(g)}S$.

It is now a straightforward task to compute the tensor R_{kij} defined in eq. (2.48). Instead of the vacuum expression (2.49) we obtain, using the above equations and the 3–dimensional Riemann tensor to eliminate the second covariant derivatives of S_k,

$$R_{kij} = \frac{2}{S^2}S_{[j}S_{i]k} + \frac{1}{S}R_{jink}^{(g)}S^n + \frac{2}{S^2 - 1 + m^2} \times$$

$$\left[\frac{S}{S^2 - 1 + m^2}S^n S_n g_{k[i}S_{j]} + S^n g_{k[j}S_{i]n} + 2\,S_{[i}S_{j]k}\right]. \qquad (9.55)$$

We now take advantage of the general expression (2.50) for the 3–dimensional Riemann tensor and obtain, again using the field equations (9.53) and (9.54),

$$R_{jink}^{(g)}S^n = \frac{2}{S}S_{[j}S_{i]k} + \frac{2}{S}S^n g_{k[i}S_{j]n} + \frac{2\,S^n S_n}{S^2 - 1 + m^2}g_{k[j}S_{i]}. \qquad (9.56)$$

Inserting this in the above expression for R_{kij} yields the result

$$R_{kij} = \left[1 - \frac{S^2}{S^2 - 1 + m^2}\right] \times$$

$$\left[\frac{4}{S^2}S_{[j}S_{i]k} + \frac{2}{S^2}g_{k[i}S_{j]n}S^n + \frac{2\,S^n S_n}{S^2 - 1 + m^2}\frac{1}{S}g_{k[j}S_{i]}\right], \qquad (9.57)$$

which replaces the corresponding vacuum expression (9.29). Now we consider $R_{kij}R^{kij}$ and use the general identity (9.33) to express the square of the first two terms in the above equation in terms of the quantities characterizing the geometry of the 2–surfaces with constant S. Performing the necessary manipulations gives

$$\frac{S^4}{8w^2} R_{kij} R^{kij} = \left[1 - \frac{S^2}{S^2 - 1 + m^2}\right]^2 \left[\overset{\circ}{H}^2 + \frac{\beta^{ij} w_i w_j}{8\,w^2} + \frac{(\Delta^{(g)} S)^2}{2\,w}\right]$$

$$+ \left[1 - \frac{S^2}{S^2 - 1 + m^2}\right]^2 \frac{S}{2\,(S^2 - 1 + m^2)} \left[\frac{w\,S}{S^2 - 1 + m^2} - 2\Delta^{(g)} S\right]$$

$$= \left[\frac{m^2 - 1}{S^2 - 1 + m^2}\right]^2 \left[\overset{\circ}{H}^2 + \frac{\beta^{ij} w_i w_j}{8w^2} + \frac{1}{2}(\frac{\Delta^{(g)} S}{w} - \frac{S}{S^2 - 1 + m^2})^2\right],$$

where $\overset{\circ}{H}{}^2 \equiv \overset{\circ}{H}_{ij} \overset{\circ}{H}{}^{ij}$. Since the term in the parenthesis vanishes as a consequence of the field equation (9.52), we finally obtain the desired result

$$\frac{S^4}{8w^2} R_{kij} R^{kij} = \left[\frac{M^2 - Q^2}{M^2 + Q^2(S^2 - 1)}\right]^2 \left[\overset{\circ}{H}_{ij} \overset{\circ}{H}{}^{ij} + \frac{\beta^{ij} w_i w_j}{8w^2}\right].$$
$$(9.58)$$

This is, up to the prefactor, exactly the same expression as we obtained in the vacuum case (see eq. (9.34)). Since this factor does not vanish in the nondegenerate case, $|Q| < M$, we again conclude that conformal flatness of (Σ, g) (i.e., $R_{kij} = 0$) implies

$$H_{ij} - \frac{1}{2} H \beta_{ij} = 0, \qquad \beta^{ij} w_{;j} = 0. \qquad (9.59)$$

The 3–geometry is thus spherically symmetric, which enables one to write the metric on Σ in the form $g = f^2 dS^2 + r^2 d\Omega^2$, where the functions f and r depend only on the radial coordinate S. The Poisson equation (9.52) for S now becomes ($f' \equiv \partial f / \partial S$)

$$\frac{(r^2 f^{-1})'}{(r^2 f^{-1})} = \frac{Q^2 S}{Q^2(S^2 - 1) + M^2},$$

which has the solution

$$f(S) = \frac{r^2(S)}{\sqrt{Q^2(S^2 - 1) + M^2}}. \qquad (9.60)$$

In addition, we consider the R_{SS}–component of eq. (9.53), which now reads

$$(\frac{r'}{r})' - \frac{f' r'}{fr} + (\frac{r'}{r})^2 = \frac{1}{2S} \left[\frac{f'}{f} + \frac{Q^2 S}{Q^2(S^2 - 1) + M^2}\right].$$

It is fulfilled for $r'(S) = Sf(S)$, with $f(S)$ given by the solution (9.60). Using the asymptotic condition $r^{-1}(S = 1) = 0$, we have $r(S) = Q^2[M - \sqrt{Q^2(S^2 - 1) + M^2}]^{-1}$. Together with $f^2 dS^2 = S^{-2} dr^2$, we finally obtain the Reissner–Nordström metric (Reissner 1916, Nordström 1918):

$$^{(4)}g = -S^2 dt^2 + S^{-2} dr^2 + r^2 d\Omega^2, \quad S^2 = 1 - \frac{2M}{r} + \frac{Q^2}{r^2}. \quad (9.61)$$

The electromagnetic potential is computed from the relation (9.51) between S and ϕ,

$$\phi^2 - 2\frac{M}{Q}\phi + 1 - S^2 = 0 \quad \Longrightarrow \quad \phi = \frac{Q}{r}. \quad (9.62)$$

This concludes the uniqueness proof for the Reissner–Nordström solution. We recall that, as in the vacuum case, the proof holds only for regular horizons; that is, if the total electric charge and the mass are subject to the inequality $|Q| < M$. Again, the reasoning does not *a priori* require a connected horizon. To summarize, we have established the following theorem (Simon 1985, Ruback 1988, Masood–ul–Alam 1992):

Theorem 9.7 *The 2–parameter Reissner–Nordström metric is the unique electrovac black hole solution with vanishing magnetic charge, regular event horizon and static, asymptotically flat domain of outer communications.*

9.4 Uniqueness of the magnetically charged Reissner–Nordström metric

In this section we consider a further, straightforward application of the methods presented above. Taking magnetic fields into account as well, we establish the nonexistence of multi black hole solutions and the uniqueness of the Reissner–Nordström metric amongst all static, asymptotically flat black hole solutions with electric *and* magnetic charge, subject to the inequality $M^2 > Q^2 + P^2$. We do so by generalizing the conformal factor introduced in eq. (9.45) in the most obvious way, that is, by replacing the square of the electric potential by the sum of the squares of the electric and the magnetic potentials (Heusler 1994). As a consequence, the expression for the conformally transformed 3–dimensional Ricci curvature becomes a difference of two positive terms, rather than a sum, as would be required in order to apply the positive mass theorem. However, using the *static* Maxwell equations, the additional term with the undesired sign is easily seen to vanish.

We start be recalling Maxwell's equations (5.25), (5.26) for invariant electromagnetic fields. In the stationary case, the expressions reduce to the two co–differential equations (8.8) for the potentials ϕ and ψ $(d\phi = E = -i_k F, \ d\psi = B = i_k * F)$. In the static case, i.e., for $\omega = 0$, we have (with $V = S^2$ and $-d^\dagger df = \Delta f = \Delta^{(g)}f + S^{-1}(df|dS))$

$$\Delta^{(g)}\phi = S^{-1}(d\phi|dS), \qquad \Delta^{(g)}\psi = S^{-1}(d\psi|dS). \qquad (9.63)$$

Instead of eq. (9.41) for $\Delta^{(g)}S$ and eq. (9.43) for $R^{(g)}$ we obtain

$$\Delta^{(g)}S = S^{-1}[(d\phi|d\phi) + (d\psi|d\psi)], \qquad (9.64)$$

$$R^{(g)} = 2\,S^{-1}\Delta^{(g)}S. \qquad (9.65)$$

In order to apply the positive mass theorem, we again consider a conformal transformation of the 3–dimensional Riemannian metric $\boldsymbol{g}$. The most obvious generalization of the transformation (9.45) of the purely electric case is

$$\Omega_\pm = \frac{1}{4}[(1 \pm S)^2 - \Lambda\overline{\Lambda}], \quad \text{where} \quad \Lambda = -\phi + i\psi. \qquad (9.66)$$

Using this expression for $\Omega_\pm$ and the field equations (9.63)-(9.65), we find instead of eq. (9.48),

$$\frac{\Omega_\pm^4}{2}\hat{R} = \frac{1}{S^2}\left[\Omega_\pm^2 \mp \Omega_\pm S + \frac{1}{4}S^2\Lambda\overline{\Lambda}\right](d\Lambda|d\overline{\Lambda}) + \frac{1}{4}\Lambda\overline{\Lambda}\,|dS|^2 +$$

$$\frac{1 - S^2 - \Lambda\overline{\Lambda}}{8\,S}(\Lambda d\overline{\Lambda} + \overline{\Lambda}d\Lambda|dS) + \frac{1}{16}[(\overline{\Lambda}d\Lambda|\overline{\Lambda}d\Lambda) + (\Lambda d\overline{\Lambda}|\Lambda d\overline{\Lambda})].$$

Since the term in the first bracket is equal to $[\frac{1}{4}(1 - S^2 - \Lambda\overline{\Lambda}]^2$, we can write the above expression as a difference of two squares. Hence, the Ricci curvature of the conformal 3–manifold $(\hat{\Sigma}, \hat{g})$ becomes

$$4^2\frac{\Omega_\pm^4}{2}\hat{R} = \left|(1 - S^2 - \Lambda\overline{\Lambda})\frac{d\Lambda}{S} + 2\Lambda\,dS\right|^2 - \left|\Lambda d\overline{\Lambda} - \overline{\Lambda}d\Lambda\right|^2, \qquad (9.67)$$

which reduces to the expression (9.49) in the purely electric case. At first glance, this expression does not look very promising, since $\hat{R}$ is not manifestly non–negative. However, if it is possible to show that the second term vanishes, then the positive mass theorem can be applied again in order to conclude that the first term also vanishes. In fact, the following proposition yields the desired result:

Proposition 9.8 *Consider a static electrovac spacetime with Killing field k. Then the Einstein–Maxwell equations imply that either the electric field, $E = -i_k F$, or the magnetic field, $B = i_k * F$, vanishes or that they are parallel with a constant factor of proportionality,*

$$B = cE, \quad \text{where} \quad c = \text{constant}. \tag{9.68}$$

Proof We start by recalling the identity (2.16) which, together with Einstein's equations and eq. (5.13), becomes

$$d\omega = -2E \wedge B. \tag{9.69}$$

Hence, in a static spacetime, $\omega = 0$ implies the existence of a function c such that the 1–forms E and B are proportional, $B = cE$ (provided that neither field vanishes). It remains to show that c is constant. The first set of Maxwell's equations, $dE = 0$, $dB = 0$, implies that the 1–forms E and dc are parallel, since

$$0 = dB = dc \wedge E + c\, dE = dc \wedge E.$$

On the other hand, the co–derivative equations (8.9) become in the static case $d^\dagger(E/S^2) = 0$ and $d^\dagger(B/S^2) = 0$, which yields

$$0 = d^\dagger\left(\frac{B}{S^2}\right) = d^\dagger\left(\frac{cE}{S^2}\right) = c\,d^\dagger\left(\frac{E}{S^2}\right) - \frac{(dc|E)}{S^2} = -\frac{(dc|E)}{S^2},$$

implying that dc is perpendicular to E. Since, by definition, E is spacelike (k is timelike), there exists no nonvanishing 1–form which is both parallel and orthogonal to E. Hence c is a constant, also implying (together with the asymptotic conditions $\phi \to 0$, $\psi \to 0$) that $\psi = c\,\phi$. □

The fact that in a static domain of outer communications E and B are proportional with a constant factor was established already by Müller zum Hagen *et al.* (1974) by different means. This shows that the second term in the expression (9.67) for $\hat{R}$ vanishes:

$$\Lambda d\overline{\Lambda} - \overline{\Lambda} d\Lambda = 2i\,(\psi\,E - \phi\,B) = 0. \tag{9.70}$$

The Ricci curvature of the conformal manifold $(\hat{\Sigma}, \hat{g})$ therefore becomes

$$\frac{\Omega_\pm^4}{2} \hat{R} = \frac{1+c^2}{(4S)^2} \left| (1 - S^2 - \phi_c^2) \, d\phi_c + 2\phi_c \, S \, dS \right|^2, \qquad (9.71)$$

where $\phi_c \equiv \sqrt{1+c^2}\,\phi$. This is, up to a constant factor, exactly the same expression as in the purely electric case (9.49), provided that ϕ is replaced by ϕ_c. The asymptotic decay conditions

$$\phi = \frac{Q}{r} + \mathcal{O}(r^{-2}), \quad \psi = \frac{P}{r} + \mathcal{O}(r^{-2}) \quad \text{as } r \to \infty, \qquad (9.72)$$

yield $c = P/Q$, where Q and P denote the electric and the magnetic charge, respectively, as measured by an observer at spacelike infinity (see eqs. (8.14), (8.15)). Asymptotic flatness requires that $S = 1 - Mr^{-1} + \mathcal{O}(r^{-2})$ and

$$\phi_c = \frac{\sqrt{Q^2 + P^2}}{r} + \mathcal{O}(r^{-2}) \quad \text{as } r \to \infty, \qquad (9.73)$$

also implying that $g = (1 + 2Mr^{-1})\,\delta + \mathcal{O}(r^{-2})$. In addition, ϕ_c fulfils the same Maxwell equations as ϕ in the purely electric case. Hence, the entire line of reasoning presented for vanishing magnetic fields can now be directly adopted, provided that the quantities ϕ and Q in the previous section are replaced by ϕ_c and $\sqrt{Q^2 + P^2}$. Assuming

$$\sqrt{Q^2 + P^2} < M, \qquad (9.74)$$

enables one to conclude that the results derived in the previous section remain true if the above replacements are made. This establishes the following generalization of theorem 9.7:

Theorem 9.9 *The 3–parameter Reissner–Nordström metric,*

$$^{(4)}g = -S^2 \, dt^2 + S^{-2} \, dr^2 + r^2 \, d\Omega^2, \qquad (9.75)$$

describes the unique electrovac black hole solution with regular event horizon and static, asymptotically flat domain of outer communications. The gravitational potential S and the electromagnetic potentials ϕ and ψ are, respectively

$$S^2 = 1 - \frac{2M}{r} + \frac{Q^2 + P^2}{r^2}, \quad \phi = \frac{Q}{r}, \quad \psi = \frac{P}{r}. \qquad (9.76)$$

It is worth pointing out that there exists no additional difficulty in proving the above uniqueness theorem, once the key observation (9.68) is established: Performing a duality rotation, the uniqueness of the 3–parameter Reissner–Nordström family is a direct consequence of this observation and the uniqueness theorem for the purely electric case.

This concludes our discussion of the uniqueness theorems for static electrovac black holes. A further application of the above methods will be discussed in chapter 12. It concerns the generalization of the static uniqueness theorem to black hole configurations with scalar fields or, more generally, nonlinear sigma models. It is also interesting to reconsider the above reasoning for non–Abelian gauge fields. As is well-known, there *do* exist black hole solutions of such theories with *vanishing* global Yang–Mills charges (see page 115 for references). The fact that these configurations are not identical with the Schwarzschild solution excludes a generalization of the Abelian uniqueness result to non–Abelian gauge fields. Considering purely electric configurations, it is nevertheless possible to gain non–Abelian uniqueness results along the lines presented above.

9.5 Multi black hole solutions

The uniqueness theorem for the Reissner–Nordström metric is restricted to configurations which are subject to the inequality $|Q| < M$ (or $\sqrt{Q^2 + P^2} < M$). Hence, the case $|Q| = M$ must be considered separately. We have already established the relation

$$S^2 - 1 = \phi^2 - 2\frac{M}{Q}\phi \qquad (9.77)$$

between the gravitational potential S and the electric potential ϕ in two ways (see sections 8.3 and 8.4). For $|Q| = M$ this reduces to the linear expression

$$S = 1 - \phi, \qquad (9.78)$$

which is valid for the *extreme* Reissner–Nordström solution. However, the fact that the latter is not covered by the uniqueness theorem raises the question of whether there exist further solutions for which the electric and the gravitational potential are subject to this simple relation. This is indeed the case:

In Newtonian gravity, a system of n arbitrarily located, charged mass points with $|q_i| = \sqrt{G}m_i$ will remain in static equilibrium. This suggests that in general relativity the gravitational attraction of n black holes can be balanced by their electrostatic repulsion, provided that for each of them $|q_i| = \sqrt{G}m_i$, where all q_i have the same sign. (Here m_i and q_i denote the asymptotically measured mass and charge of spacetime in the absence of all but the ith black hole.) In fact, Papapetrou (1945) and Majumdar (1947) independently found a solution of the Einstein–Maxwell equations of this kind, representing n arbitrarily arranged extreme Reissner–Nordström black holes. In order to derive the Papapetrou–Majumdar metric, we use $\phi = 1 - S$ in the static Einstein–Maxwell equations (9.41)-(9.44). Both the Maxwell equation (9.44) and the Poisson equation (9.41) for the gravitational potential then become,

$$\Delta^{(g)} S = S^{-1} (dS|dS)^{(g)}. \qquad (9.79)$$

In addition, one has to consider the Einstein equations on Σ. This is most easily done in the projection metric $\boldsymbol{P}$ introduced in eq. (2.19). In the static case (where $K = k = -S^2 dt$) the latter is given by $(N = -S^2)$

$$^{(4)}\boldsymbol{g} = -S^2\, dt \otimes dt + \frac{1}{S^2}\, \boldsymbol{P}. \qquad (9.80)$$

The remaining Einstein equations are now obtained from the projection formula (2.20). In the electromagnetic case, this yields the expression (5.15), which for $\omega = 0$ and $B = 0$ becomes

$$\boldsymbol{R}^{(P)} = \frac{2}{S^2}\, dS \otimes dS - \frac{2}{S^2}\, d\phi \otimes d\phi. \qquad (9.81)$$

Now inserting the relation (9.78) between ϕ and S immediately yields $\boldsymbol{R}^{(P)} = 0$, which shows that $(\Sigma, \boldsymbol{P})$ is a *flat* 3–manifold. Hence, the remaining Einstein equations are fulfilled for $\boldsymbol{P} = \boldsymbol{\delta}$ and the entire set of static electrovac equations reduces to

$$^{(4)}\boldsymbol{g} = -S^2\, dt \otimes dt + S^{-2}\, \boldsymbol{\delta}, \qquad (9.82)$$

$$\Delta^{(\delta)} S^{-1} = 0, \qquad \phi = 1 - S. \qquad (9.83)$$

(Note that $\boldsymbol{g} = S^{-2}\boldsymbol{\delta}$ yields $\Delta^{(g)} S = S^2 \Delta^{(\delta)} S - S(dS|dS)^{(\delta)}$ and thus $\Delta^{(\delta)} S^{-1} = 0$.)

The simplest nontrivial solution of the flat Poisson equation corresponds to a linear combination of n monopole sources m_i located at arbitrary points $\underline{x}_i$,

$$S^{-1}(\underline{x}) = 1 + \sum_i^n \frac{m_i}{|\underline{x} - \underline{x}_i|}, \qquad (9.84)$$

(with $S \to 1$ as $|\underline{x}| \to \infty$). If the sources are widely separated, m_i can be interpreted as the mass enclosed by a distant sphere centered at $\underline{x}_i$. The total mass and charge are equal, since

$$Q = -\frac{1}{4\pi} \int_{S_\infty^2} *F = -\frac{1}{4\pi} \int_{S_\infty^2} *(k \wedge \frac{E}{V}) = \frac{1}{4\pi} \int_{S_\infty^2} *(k \wedge \frac{dS}{V})$$

$$= \frac{1}{8\pi S_\infty} \int_{S_\infty^2} *(k \wedge \frac{dV}{V}) = -\frac{1}{8\pi} \int_{S_\infty^2} *dk = M, \qquad (9.85)$$

which holds for every solution with $S = 1 - \phi$. (Here we have used the general relations (8.16) and (8.17) with $V \equiv S^2$ and $E = d\phi = -dS$.) For $n = 1$, the Papapetrou–Majumdar solution reduces to the extreme Reissner–Nordström solution.

Hartle and Hawking (1972) have given arguments which indicate that the Papapetrou–Majumdar geometries obtained from other solutions than (9.84) of the Poisson equation $\Delta^{(\delta)} S^{-1} = 0$ exhibit naked singularities. The above solution describes a *regular* black hole spacetime with n disconnected components of the event horizon, represented by the surfaces $|\underline{x}_i| = 0$. All real singularities are "hidden" behind these null surfaces (Hartle and Hawking 1972). This is most easily seen by analyzing the geometry of two sources,

$$S^{-1} = 1 + \frac{m_1}{r} + \frac{m_2}{\sqrt{r^2 + a^2 - 2ar\cos\theta}}. \qquad (9.86)$$

(Here the first source is placed at the origin of a polar coordinate system with the second source lying on the z–axis at the distance a.) As is seen from the metric

$$^{(4)}g = -S^2 \, dt^2 + S^{-2} [\, dr^2 + r^2 \, d\Omega^2 \,], \qquad (9.87)$$

the area of a small sphere centered at the origin of the above coordinate system remains finite as $r \to 0$, since $4\pi S^{-2} r^2 \to 4\pi m_1^2$. This indicates that $r = 0$ is a surface, rather than a point. In fact, $r = 0$ is a regular null surface. In order to make this evident, one considers the coordinate transformation (see Chandrasekhar 1983)

$$\tau = t - f(r), \quad \text{where} \quad \frac{df}{dr} \equiv (1 + \frac{m_2}{a} + \frac{m_1}{r})^2.$$

Using the asymptotic behavior of S^{-2} and df/dr, one has $(S^{-1} - \sqrt{df/dr}) = a^{-2}m_2 \cos\theta\, r + \mathcal{O}(r^2)$ as $r \to 0$, from which one finds

$$^{(4)}\boldsymbol{g} \to -(\frac{r^2}{m_1^2}\, d\tau^2 + 2\, d\tau\, dr) + 4\frac{m_1 m_2 \cos\theta}{a^2}\, dr^2 + m_1^2\, d\Omega^2 ,$$

as $r \to 0$. Hence, the metric remains regular as $r \to 0$ and it is easy to see that $r = 0$ is indeed a null surface with area $4\pi m_1^2$.

In order to extend the manifold to negative values of r, one considers the transformation $r \to r'$ and the solution $(S')^{-1}$, defined by

$$r \to r' = -r, \tag{9.88}$$

$$(S')^{-1} = 1 - \frac{m_1}{r'} + \frac{m_2}{\sqrt{(r')^2 + a^2 + 2ar' \cos\theta}}. \tag{9.89}$$

The metric $^{(4)}\boldsymbol{g}' = -(S')^2 dt^2 + (S')^{-2}(d(r')^2 + (r')^2 d\Omega^2)$ becomes singular at the surface $(S')^{-1} = 0$. As is seen from the diverging curvature invariants, this is now a real singularity. Again computing the area, one finds that $(S')^{-1} = 0$ is not a surface, but a point singularity, which is "hidden" behind the null–surface $r = 0$. We refer the reader to Hartle and Hawking (1972) for a detailed description of this spacetime.

The uniqueness theorem for the Papapetrou–Majumdar metric has not yet been established conclusively. Recent progress has been achieved by Chruściel and Nadirashvili (1995), who were able to settle the uniqueness of the *standard* Papapetrou–Majumdar spacetime amongst the nonsingular (in an appropriate sense) electrovac black hole spacetimes of the Papapetrou–Majumdar form (see Perjes 1971, Israel and Wilson 1972, Hartle and Hawking 1972). A further step towards a complete classification of the static, asymptotically flat electrovac spacetimes could be achieved by proving the equality $M = |Q|$, which is presumed to hold if the horizon has degenerate components. It is possible that the methods presented in section 8.3 will provide a useful tool in settling this problem. Partial answers are known for the case where all components of the Killing horizon are assumed to be degenerate and all "horizon charges" are required to have the same sign (Heusler 1995c).

In the next chapter we shall discuss the uniqueness of the stationary Kerr–Newman family with parameters M, J and Q, subject to the inequality $M^2 > (J/M)^2 + Q^2$. If this restriction is dropped, one may again look for axisymmetric spacetimes containing more than a single black hole. In fact, Israel and Wilson (1972) found stationary generalizations of the Papapetrou–Majumdar metric. However, these solutions share certain unpleasant properties with NUT spacetime (Newman *et al.* 1963; see also Brill 1964, Misner 1965): The work of Hartle and Hawking (1972) and Chruściel and Nadirashvili (1995) strongly suggests that the solutions obtained by the Israel–Wilson technique are either not asymptotically Euclidean or exhibit naked singularities. Hence, it is most likely that the static Papapetrou–Majumdar spacetime is the only asymptotically flat, regular multi black hole solution of the Einstein–Maxwell equations. The reader is referred to Weinstein (1992, 1994a, 1994b) for a detailed treatment of axisymmetric multi black hole configurations.

10

Uniqueness theorems for rotating holes

In this chapter we present the uniqueness theorem for the Kerr–Newman metric. The latter describes the only asymptotically flat, stationary and axisymmetric electrovac black hole solution with regular event horizon. The proof of this fact involves the following steps: First, one has to establish circularity of the domain of outer communications as a consequence of the symmetry properties of the electromagnetic field. The Einstein–Maxwell equations are then reduced to a 2–dimensional elliptic boundary–value problem for the complex Ernst potentials E and Λ. One then takes advantage of the symmetries of the Ernst equations to derive a divergence identity for the difference of two solutions. Since the boundary and regularity conditions are completely parametrized in terms of the total mass, angular momentum and charge, Stokes' theorem finally yields the desired result.

The chapter is organized as follows: The first section gives a short outline of the reasoning. In the second section we parametrize the Ernst potentials in terms of the hermitian matrix Φ, describing the nonlinear sigma model on the symmetric space $G/H = SU(1,2)/S(U(1) \times U(2))$ (or $G/H = SU(1,1)/U(1)$ in the vacuum case). We then establish the variational equation for Φ and derive an identity for the difference of two solutions to this equation. Evaluating the expressions in a circular spacetime, we obtain the Mazur identity (Mazur 1982) in the third section. This identity - or a related identity found by Bunting in 1983 - must be considered the key to the uniqueness theorems for rotating black holes. In the vacuum case, it reduces to the famous identity obtained by Robinson in 1975, which will be discussed in the fourth section. In order to understand the existence of the positive and hermitian matrix Φ, we conclude this chapter with some comments on the relationship between harmonic maps and sigma models.

10.1 Outline of the reasoning

We start by briefly describing the main issues involved in the proof of the uniqueness theorem.

Concerning the circularity problem, we refer to theorem 5.7, which establishes the integrability conditions from the symmetry properties of the electromagnetic field with respect to the stationary and the axial Killing field.

Once circularity is established, the field equations can be reduced to a 2–dimensional boundary–value problem (Carter 1971, 1973c). This is achieved in Weyl coordinates, using the fact that $\rho = \sqrt{-\sigma}$ is a harmonic function (with respect to the 2–dimensional metric of the Riemannian manifold orthogonal to the Killing orbits) for both the vacuum and the electrovac case (see Carter 1973a, Weinstein 1990 and section 3.4). The Einstein–Maxwell equations in terms of the two complex Ernst potentials E and Λ were derived in section 5.5. A very important feature of the 2–dimensional boundary–value problem lies in the Lagrangian structure of the Ernst equations (Carter 1973a). As we have emphasized several times, the effective Lagrangian turns out to be positive definite only if the Ernst system is based on the axial Killing field.

The last - and most difficult - part of the uniqueness proof is to show that two solutions of the boundary–value problem are equal if they are subject to the same set of boundary and regularity conditions. In the case of *infinitesimally* different solutions, Carter (1971) solved this problem for the vacuum case by means of a divergence identity. Subsequently, Robinson (1974) found a complicated identity which enabled him to extend Carter's result to electrovac spacetimes.

Considering two arbitrary, *not* necessarily infinitesimally neighboring solutions of the Ernst equations, Robinson (1975) also succeeded in proving the uniqueness of the vacuum Kerr metric. The fact that the *linearized* electrovac identity was very complicated compared with the linearized vacuum identity, and the already complicated nature of the *full vacuum* identity, dashed the hope to find the *full electrovac* identity by using trial and error methods:

> If such a generalization exists it will certainly be extremely complicated ... . I feel strongly that there must be a deep but essentially simple mathematical reason why the identities found so far should

exist, and I would conjecture that the generalization required to tie up the problem completely will not be constructed *after* the discovery of such an underlying explanation, which would presumably show one how to construct the required identities *directly* (without recourse to the algebraic trial and error method to which Robinson and I were obliged to resort). *Carter (1979)*

In fact, Carter's prediction was shown to be true when Mazur (1982, 1984b) and Bunting (1983) independently succeeded in deriving the desired divergence identities in a systematic way: Their different methods, which prove the uniqueness of the Kerr–Newman metric, are based on the circumstance that the effective Lagrangian is a quadratic expression in terms of the exterior derivatives of the Ernst potentials. Bunting's approach, applying to a general class of harmonic mappings between Riemannian manifolds, yields an identity which enables one to establish the uniqueness of a harmonic map if the target manifold has negative curvature. Since we shall not go into the details of this construction, we refer the reader to Carter (1985) for a discussion of Bunting's method.

The Mazur identity, which we discuss in this chapter, is based on the observation that the Ernst equations describe a nonlinear sigma model on the symmetric space G/H, where G is a connected Lie group and H is a maximal compact subgroup of G. In the vacuum case, $G/H = SU(1,1)/U(1)$, whereas $G/H = SU(1,2)/S(U(1) \times U(2))$ if electromagnetic fields are also taken into account. In both situations it is possible to establish the identity

$$- \tilde{d}^\dagger \left[\rho \, d(\mathrm{Tr}\, \Psi) \right] + \mathrm{Tr} \left\{ \Phi_{(2)} \, \tilde{d}^\dagger (\rho \, \triangle J) \, \Phi_{(1)}^{-1} \right\}$$
$$= \rho \, \mathrm{Tr} \left(\Phi_{(1)}^{-1} \triangle J^\dagger \, | \, \Phi_{(2)} \triangle J \right)^{\sim}, \qquad (10.1)$$

where $\Phi_{(A)}$ $(A = 1, 2)$ are two G-valued fields, $J_{(A)} \equiv \Phi_{(A)}^{-1} d\Phi_{(A)}$, $\triangle J \equiv J_{(2)} - J_{(1)}$ and $\Psi \equiv \Phi_{(2)} \Phi_{(1)}^{-1} - \mathbb{1}$. (Throughout this chapter, quantities with a tilde refer to the metric (5.87) of the 2–dimensional Riemannian manifold $(\Gamma, \tilde{\gamma})$; e.g., $(\alpha|\beta)^{\sim} \equiv \alpha_i \beta_j \tilde{\gamma}^{ij}$.) Since the second term vanishes for two *solutions* of the Ernst equations, Stokes' theorem yields a relation between the boundary integral of $\rho \, d(\mathrm{Tr}\, \Psi)$ and the integral of the r.h.s. of the above identity. Using the fact that Φ is a positive and hermitian matrix whenever G/H

is of the form $SU(p,q)/S(U(p) \times U(q))$, as well as the circumstance that $\tilde{\gamma}$ is a *Riemannian* metric, one finds that the integrand on the r.h.s. is non–negative. The Mazur identity implies then that two solutions, $\Phi_{(1)}$ and $\Phi_{(2)}$, for which $\rho\, d(\operatorname{Tr}\Psi)$ vanishes at the boundary, are in fact equal.

10.2 The Ernst system and the Kinnersley group

The purpose of this section is to rewrite the action (5.42) in terms of a hermitian $SU(1,2)$–valued matrix Φ. We then formulate the Ernst equations for the potentials E and Λ associated with a Killing field K in terms of Φ. An identity for the Laplacian of the relative difference of two configurations, $\Phi_{(1)}$ and $\Phi_{(2)}$, is established at the end of this section.

We point out that the results derived in this section do not depend on the spacetime metric. The only property of spacetime which is used is the existence of *one* Killing field, K, say. If K is spacelike, then the electrovac action describes a harmonic mapping into the symmetric space $SU(1,2)/S(U(1) \times U(2))$. (If no electromagnetic fields are taken into account one obtains, instead, a mapping with target manifold $SU(1,1)/U(1)$.) If the Ernst equations are formulated on the basis of a timelike Killing field, then the target manifold becomes $SU(1,2)/S(U(1) \times U(1,1))$.

In the next section we shall evaluate the results for a circular spacetime with $K = m$. In this case Φ is positive and describes a mapping from the 2–dimensional Riemannian manifold $(\Gamma, \tilde{\gamma})$ into $SU(1,2)/S(U(1) \times U(2))$.

The following three propositions are the key to the Mazur identity. They are valid for an arbitrary spacetime metric $^{(4)}g$ admitting a Killing field K. (We use the notation $|\alpha|^2 = {}^{(4)}g^{\mu\nu}\alpha_\mu\bar{\alpha}_\nu$ for the norm of a complex 1–form α with respect to $^{(4)}g$.)

Proposition 10.1 *Let* E *and* Λ *be the Ernst potentials associated with a Killing field K with norm $N = (K|K) = -Re(\mathrm{E}) - \Lambda\bar{\Lambda}$. Then the Lagrangian*

$$\mathcal{L} = \frac{|d\mathrm{E} + 2\bar{\Lambda}d\Lambda|^2}{N^2} + 4\frac{|d\Lambda|^2}{N} \tag{10.2}$$

can be written in the form

$$\mathcal{L}[\Phi] = \frac{1}{2}\operatorname{Tr}(J|J)\,, \quad with \;\; J = \Phi^{-1}d\Phi\,, \qquad (10.3)$$

where the hermitian $SU(1,2)$ *matrix* Φ *is defined by*

$$\Phi_{ab} = \eta_{ab} + 2\operatorname{sign}(N)\,\overline{v}_a v_b\,. \qquad (10.4)$$

Here $\eta = \operatorname{diag}(-1,+1,+1)$, $\operatorname{sign}(N) = N/|N|$ *and* v *is the Kinnersley vector,*

$$(v_0\,,\,v_1\,,\,v_2) = \frac{1}{2\sqrt{|N|}}\,(\,\mathrm{E}-1\,,\,\mathrm{E}+1\,,\,2\Lambda\,)\,. \qquad (10.5)$$

Proof The reasons for the existence of the matrix Φ and the sigma model structure of the Ernst system are explained in the last section of this chapter. For the moment, we content ourselves with a verification of the above assertions: We first note that the definition (10.5) implies the normalization

$$v_a\overline{v}^a \equiv \eta^{ab}v_a\overline{v}_b = -\operatorname{sign}(N)\,. \qquad (10.6)$$

A straightforward calculation shows that the Lagrangian (10.2) now assumes the form

$$\mathcal{L} = 4\,[\,\operatorname{sign}(N)\,(dv_a\,|\,d\overline{v}^a) + (v_a d\overline{v}^a\,|\,\overline{v}^b dv_b)\,]\,. \qquad (10.7)$$

Differentiating the definition (10.4) and using the inverse matrix, $\Phi^{ab} = \eta^{ab} + 2\operatorname{sign}(N)\,\overline{v}^a v^b$, we find

$$J^a{}_b = 2\operatorname{sign}(N)\,(\,v_b d\overline{v}^a - \overline{v}^a dv_b\,) + 4\,\overline{v}^a v_b\,v^c d\overline{v}_c\,, \qquad (10.8)$$

which we use to compute $\operatorname{Tr}(J|J) \equiv (J^a{}_b\,|\,J^b{}_a)$. Also using the consequence $\overline{v}^a dv_a + v_a d\overline{v}^a = 0$ of the normalization (10.6), establishes that the Lagrangians (10.2), (10.3) and (10.7) are equal. $\square$

Proposition 10.2 *The variational equation for* $\int \mathcal{L}[\Phi] * 1$ *with Lagrangian (10.3) is*

$$d * J = 0\,. \qquad (10.9)$$

Proof Since $\delta\Phi^{-1} = -\Phi^{-1}\delta\Phi\Phi^{-1}$ and $d\Phi^{-1} = -J\Phi^{-1}$, we obtain

$$\delta J = J\,\Phi^{-1}\delta\Phi - \Phi^{-1}\delta\Phi\,J + d\,(\Phi^{-1}\delta\Phi)\,.$$

Performing the variation of the action with respect to Φ and integrating by parts now yields

$$\frac{1}{2}\,\delta\operatorname{Tr}(J\wedge *J) \doteq -\operatorname{Tr}\{\,\Phi^{-1}\delta\Phi\,([*J,J] + d * J)\,\}\,,$$

where '$\doteq$' stands for equal up to an exact differential, and where we have used $\mathrm{Tr}\{[J, \Phi^{-1}\delta\Phi] \wedge *J\} = \mathrm{Tr}\{[*J, J]\,\Phi^{-1}\delta\Phi\}$. Since this term vanishes, we obtain the desired result. $\qquad\square$

We are now ready to derive the basic divergence identity. It relates the Laplacian of the difference of two configurations, $\Phi_{(1)}$ and $\Phi_{(2)}$, say, to a quadratic expression in $J_{(2)} - J_{(1)}$. As we shall argue in the next section, this expression becomes positive in a stationary and axisymmetric spacetime, provided that the $\Phi_{(A)}$ are formed with respect to the Ernst potentials associated with the axial Killing field.

Proposition 10.3 *Let $\Phi_{(A)} = \Phi^{\dagger}_{(A)}$ be hermitian matrices and let $J_{(A)} = \Phi^{-1}_{(A)}\,d\Phi_{(A)}$ $(A = 1, 2)$. Then*

$$-d^{\dagger}d\,\mathrm{Tr}\,\Psi = -\mathrm{Tr}\,\{\,\Phi_{(2)}\,d^{\dagger}(\Delta J)\,\Phi^{-1}_{(1)}\,\} + \mathrm{Tr}\,(\Phi^{-1}_{(1)}\,\Delta J^{\dagger}\,|\,\Phi_{(2)}\,\Delta J)\,,$$
$$(10.10)$$

where

$$\Psi \equiv \Phi_{(2)}\Phi^{-1}_{(1)} - \mathbb{1}\,, \qquad \Delta J \equiv J_{(2)} - J_{(1)}\,. \qquad (10.11)$$

If, in addition, the $\Phi_{(A)}$ are solutions to the variational equations (10.9) for $\int *\mathcal{L}[\Phi]$, then

$$-d^{\dagger}d\,\mathrm{Tr}\,\Psi = \mathrm{Tr}\,(\Phi^{-1}_{(1)}\,\Delta J^{\dagger}\,|\,\Phi_{(2)}\,\Delta J) = \mathrm{Tr}\,(\Delta J^{\dagger}\,|\,d\Psi)\,. \quad (10.12)$$

Proof As an immediate consequence of the definitions we obtain $d\Psi = \Phi_{(2)}\,\Delta J\,\Phi^{-1}_{(1)}$. Since $d^{\dagger}(f\alpha) = -(df\,|\,\alpha) + f d^{\dagger}\alpha$ for arbitrary 1–forms α and functions f, we find

$$-d^{\dagger}d\,\Psi = -\ \Phi_{(2)}\,d^{\dagger}(\Delta J)\,\Phi^{-1}_{(1)}$$
$$+\ (d\Phi_{(2)}\,|\,\Delta J)\,\Phi^{-1}_{(1)} + \Phi_{(2)}\,(\Delta J\,|\,d\Phi^{-1}_{(1)})\,.$$

In order to simplify this expression, we note that $\Phi = \Phi^{\dagger}$ implies $d\Phi = (\Phi^{-1}d\Phi)^{\dagger}\Phi = J^{\dagger}\Phi$ and thus $d\Phi^{-1} = -\Phi^{-1}J^{\dagger}$. Hence, the trace of the second line becomes

$$\mathrm{Tr}\,\{\,(\Phi^{-1}_{(1)}\,J^{\dagger}_{2}\,\Phi_{(2)}\,|\,\Delta J) - (\Delta J\,|\,\Phi^{-1}_{(1)}\,J^{\dagger}_{1}\,\Phi_{(2)})\,\}$$
$$=\ \mathrm{Tr}\,(\Phi^{-1}_{(1)}\,\Delta J^{\dagger}\,|\,\Phi_{(2)}\,\Delta J) = \mathrm{Tr}\,(\Delta J^{\dagger}\,|\,d\Psi)\,,$$

which completes the proof of the identity (10.10). If the $\Phi_{(A)}$ are subject to the equations $d*J_{(A)} = 0$, then $d^{\dagger}(\Delta J)$ vanishes, implying eq. (10.12). $\qquad\square$

10.3 The uniqueness proof

We are now in a position to prove the uniqueness theorem for the Kerr–Newman metric. According to proposition 5.9, the metric of an asymptotically flat, stationary and axisymmetric electrovac spacetime is parametrized in terms of the three functions X, A and h,

$$^{(4)}g = -\frac{\rho^2}{X}\,dt^2 + X\,(d\varphi + A\,dt)^2 + \frac{1}{X}e^{2h}\mu^2(x^2 - y^2)\,\tilde{\gamma}\,, \quad (10.13)$$

with $\tilde{\gamma}$ according to eq. (5.87) (see eq. (10.15)). Once the complex potentials E and Λ have been computed, the functions A and h are obtained by quadrature (see eqs. (5.77) and (5.78)).

Let us now evaluate the expressions obtained in the previous section for the circular metric (10.13). We consider the Ernst potentials and the corresponding matrices Φ which are constructed on the basis of the axial Killing field, i.e., we set $K = m$, $N = X$, sign$(N) = +1$. We recall that for stationary and axisymmetric 1–forms α one has

$$d^\dagger \alpha = \frac{1}{\rho}\tilde{d}^\dagger(\rho\,\alpha)\,, \quad (10.14)$$

where quantities furnished with a tilde refer to the metric

$$\tilde{\gamma} = \left(\frac{dx^2}{x^2 - 1} + \frac{dy^2}{1 - y^2}\right). \quad (10.15)$$

We now obtain the following corollary to the above propositions:

Corollary 10.4 *Consider a circular spacetime with metric as given in eq. (10.13). Then the Lagrangian (10.2) describes a nonlinear sigma model on the symmetric space $SU(1,2)/S(U(1) \times U(2))$ with Riemannian base manifold $(\Gamma, \tilde{\gamma})$. The matrix Φ - defined in terms of the Ernst potentials associated with the axial Killing field by eqs. (10.4), (10.5) - is positive and hermitian and subject to the Ernst equations*

$$\tilde{d}\tilde{*}\,(\rho\,J) = 0\,, \quad i.e., \quad \tilde{\nabla}_i(\rho\,J^i) = 0\,. \quad (10.16)$$

The identity (10.10) becomes the Mazur identity (10.1) which, after integration over $S \subset \Gamma$, yields

$$\int_{\partial S} \rho * d\,(\mathrm{Tr}\,\Psi) = \int_S \rho\,\mathrm{Tr}\,(\Phi_{(1)}^{-1}\Delta J^\dagger \,|\, \Phi_{(2)}\Delta J)\tilde{*}\tilde{\eta} \geq 0\,, \quad (10.17)$$

for two solutions $\Phi_{(1)}$ and $\Phi_{(2)}$ of eq. (10.16).

The only points in the above corollary which need further explanations are the assertions that Φ is positive and that the integrand on the r.h.s. of eq. (10.17) is non–negative: The matrix Φ is positive since the embedding of the symmetric space G/H in G can be represented in the form $gg^\dagger$ if $G/H = SU(p,q)/S(U(p) \times U(q))$. We refer the reader to the last section of this chapter for a brief discussion of this point. In order to obtain eq. (10.17), we have used $\tilde{d}^\dagger = -\tilde{*}d\tilde{*}$ and $\tilde{*}^2 = (-1)^p$ for the co–derivative and the Hodge dual with respect to $\tilde{\gamma}$. The semi–definiteness of the integrand on the r.h.s. is a consequence of $\rho \geq 0$, the fact that $\tilde{\gamma}$ is a Riemannian metric, ΔJ is spacelike and $\Phi_{(A)} = g_{(A)} g_{(A)}^\dagger$:

$$\mathrm{Tr}\,(\Phi_{(1)}^{-1}\Delta J^\dagger \,|\, \Phi_{(2)}\Delta J)^{\tilde{}} = \mathrm{Tr}\,(\mathcal{M}\,|\,\mathcal{M}^\dagger)^{\tilde{}} \geq 0 \,,$$

where $\mathcal{M} \equiv g_{(1)}^{-1} \Delta J^\dagger g_{(2)}$. We now obtain the following uniqueness result:

Corollary 10.5 *Let $\Phi_{(1)}$ and $\Phi_{(2)}$ denote two solutions of the sigma model equations (10.16) on $SU(p,q)/S(U(p) \times U(q))$ with 2–dimensional Riemannian base manifold $(\Gamma, \tilde{\gamma})$. Let $\rho \geq 0$ and $\rho\, d(\mathrm{Tr}\ \Psi) = 0$ on the boundary ∂S of $S \subset \Gamma$, where $\Psi \equiv \Phi_{(2)}\Phi_{(1)}^{-1} - \mathbb{1}$. Then $\Phi_{(1)} = \Phi_{(2)}$ in all of S, provided that $\Phi_{(1)}(p) = \Phi_{(2)}(p)$ for at least one point $p \in S$.*

In order to apply this result to stationary and axisymmetric black hole solutions, we must convince ourselves that the boundary and regularity conditions (5.129), (5.130) and (5.132) imply that $\rho\, d(\mathrm{Tr}\ \Psi)$ vanishes on ∂S: Since $\rho = 0$ on the horizon and on the rotation axis, it remains to be verified that $d(\mathrm{Tr}\ \Psi)$ is bounded for $x = 1$ and for $|y| = 1$. In addition, we have to establish that $\rho\, d(\mathrm{Tr}\ \Psi) \to 0$ as $x \to \infty$. Using the parametrization (10.4) for Φ in terms of v and the definition (10.5) for v, we find

$$\begin{aligned} \mathrm{Tr}\ \Psi &= \mathrm{Tr}\,(\Phi_{(2)} \Phi_{(1)}^{-1}) - (1+q) = 4\,(|\bar{v}_{(2)\,a}\, v_{(1)}^a|^2 - 1) \\ &= \frac{1}{X_{(1)} X_{(2)}} \,(|E_{(1)} + \overline{E}_{(2)} + 2\,\Lambda_{(1)}\overline{\Lambda}_{(2)}|^2 - 4\,X_{(1)}X_{(2)}) \,. \end{aligned}$$

Since $E = -X - \Lambda\overline{\Lambda} + iY$, it is now easy to express $\mathrm{Tr}\ \Psi$ in terms of the potentials X, Y and Λ. Performing the necessary manipulations gives

$$\text{Tr}\Psi = \frac{(\triangle X)^2 + |\triangle \Lambda|^2(|\triangle \Lambda|^2 + 2\Sigma X) + [\triangle Y + \text{Im}(\triangle \Lambda \Sigma \overline{\Lambda})]^2}{\Pi\, X},$$

(10.18)

where $\Pi f \equiv f_{(2)} \cdot f_{(1)}$, $\Sigma f \equiv f_{(2)} + f_{(1)}$ and $\triangle f \equiv f_{(2)} - f_{(1)}$.

We are now able to verify that $\rho\, d(\text{Tr}\,\Psi)$ vanishes on all parts of the boundary ∂S; that is, (i) for $x = \infty$, (ii) on the horizon $(x = 1)$, and (iii) on the northern and southern part of the rotation axis $(y = \pm 1)$.

(i) The asymptotic behavior of X, Y and Λ is given in terms of the total mass M, the angular momentum J and the electric charge Q by eqs. (5.129). Hence, as $x \to \infty$, the leading contribution in the numerator of eq. (10.18) is $(\triangle X)^2 = \mathcal{O}(x^2)$, whereas the leading term in the denominator is $(1 - y^2)^2 \mu_{(1)}^2 \mu_{(2)}^2 x^4$. (Note that $\mu_{(1)} = \mu_{(2)} = M^2 - (J/M)^2 - Q^2$ since $\Phi_{(1)}$ and $\Phi_{(2)}$ are solutions with the same set of asymptotic "charges".) Now using $\rho = \mathcal{O}(x)$ we find $\rho\, d(\text{Tr}\,\Psi) \to 0$ as $x \to \infty$.

(ii) On the horizon, X, Y and Λ are well–behaved functions with $X = \mathcal{O}(1)$ and $X^{-1} = \mathcal{O}(1)$ (see eqs. (5.132)). This implies that $d(\text{Tr}\,\Psi)$ remains finite for $x = 1$ which, together with $\rho = 0$, enables one to conclude that $\rho\, d(\text{Tr}\,\Psi)$ vanishes on the horizon.

(iii) The most difficult part of the boundary to deal with is $y = \pm 1$, i.e., the rotation axis, where $X = (m|m)$ vanishes. Requiring that the derivatives of the potentials in the vicinity of the axis are subject to the regularity conditions (5.130) enables one to show that X, Y and Λ are completely determined by the asymptotic conditions (5.129) on the *entire* axis; $X = 0$, $Y = \pm 4J$ and $\Lambda = \pm iQ$. The differences $\triangle X$, $\triangle Y$ and $\triangle \Lambda$ are therefore of $\mathcal{O}(1 - y^2)$ as $y \to \pm 1$, as is X. This shows that $d(\text{Tr}\,\Psi)$ remains finite on the axis, implying that $\rho\, d(\text{Tr}\,\Psi)$ vanishes for $y = \pm 1$.

By virtue of eq. (10.17), $\rho\, d(\text{Tr}\,\Psi) = 0$ implies that $\triangle J$ vanishes in the semi–strip $S = \{(x, y) \,|\, x \geq 1, |y| \leq 1\}$. Since $d\Psi = \Phi_{(2)} \triangle J\, \Phi_{(1)}^{-1}$, this shows that Ψ is a constant matrix. Considering two solutions with the same set of asymptotic charges, one has $\Psi \to 0$ as $x \to \infty$. This eventually yields the conclusion $\Psi = 0$, i.e., $\Phi_{(1)} = \Phi_{(2)}$.

Since the remaining metric functions A and h are uniquely determined by the Ernst potentials E and Λ (i.e., by Φ) and the requirement of asymptotic flatness (see eqs. (5.77), (5.78)), they must also be identical. This concludes the proof of the following uniqueness theorem, due to Robinson (1975) in the vacuum case and to Mazur (1982) and Bunting (1983) if electromagnetic fields are taken into account as well:

Theorem 10.6 *The Kerr–Newman metric (5.114) with parameters $m = M$, $a = J/M$, Q and electromagnetic field (5.120) is the only electrovac black hole solution with $M^2 > a^2 + Q^2$, vanishing magnetic charge, regular event horizon and stationary and axisymmetric, asymptotically flat domain of outer communications.*

The extension of the theorem to configurations with nonvanishing magnetic charge poses no additional difficulties. The 4–parameter Kerr–Newman metric is obtained from the construction presented in section 5.6 if the parameter λ_0 in eq. (5.109) is not restricted to real values.

10.4 The Robinson identity

In 1975 Robinson was able to establish the uniqueness of the vacuum Kerr family (with $M^2 > a^2$) amongst all stationary and axisymmetric black hole solutions with nondegenerate event horizon. The key to his proof is an identity which relates the covariant divergence of a suitably constructed quantity to a sum of manifestly non–negative terms - plus a number of additional contributions which vanish if the Ernst equations are fulfilled. In terms of the real potentials X and Y ($\mathbf{E} = -X + iY$), the Ernst equations assume the form (see eq. (4.16))

$$e(X,Y) + i f(X,Y) = 0,\qquad\qquad (10.19)$$

where

$$e(X,Y) \;\equiv\; \tilde{\nabla}\cdot(\rho X^{-2}\tilde{\nabla}X) + \frac{\rho}{X^3}(|\tilde{\nabla}X|^2 + |\tilde{\nabla}Y|^2)$$

$$f(X,Y) \;\equiv\; \tilde{\nabla}\cdot(\rho X^{-2}\,\tilde{\nabla}Y),\qquad\qquad (10.20)$$

and $\tilde{\nabla}$ denotes the covariant derivative with respect to the 2–dimensional Riemannian metric $\tilde{\gamma}$ (10.15). In terms of these functions, Robinson succeeded in "deriving" the following identity:

$$\tilde{\nabla} \cdot \left[\rho \tilde{\nabla} \left(\frac{(\Delta X)^2 + (\Delta Y)^2}{\Pi X} \right) \right] + 2 \frac{\Delta Y}{\Pi X} \left[X_{(1)}^2 f_{(1)} - X_{(2)}^2 f_{(2)} \right]$$

$$+ \frac{e_{(1)}}{X_{(2)}} \left[(\Delta Y)^2 + \Delta X \, \Sigma X \right] + \frac{e_{(2)}}{X_{(1)}} \left[(\Delta Y)^2 - \Delta X \, \Sigma X \right]$$

$$= \frac{\rho}{\Pi X} \left[\frac{\Delta Y}{X_{(1)}} \tilde{\nabla} Y_{(1)} + \frac{X_{(1)}}{X_{(2)}} \tilde{\nabla} X_{(2)} - \tilde{\nabla} X_{(1)} \right]^2$$

$$+ \frac{\rho}{\Pi X} \left[\frac{\Delta Y}{X_{(2)}} \tilde{\nabla} Y_{(2)} - \frac{X_{(2)}}{X_{(1)}} \tilde{\nabla} X_{(1)} + \tilde{\nabla} X_{(2)} \right]^2$$

$$+ \frac{\rho}{2 \, \Pi X} \left[\left(\frac{\tilde{\nabla} Y_{(2)}}{X_{(2)}} - \frac{\tilde{\nabla} Y_{(1)}}{X_{(1)}} \right) \Sigma X - \left(\frac{\tilde{\nabla} X_{(2)}}{X_{(2)}} + \frac{\tilde{\nabla} X_{(1)}}{X_{(1)}} \right) \Delta Y \right]^2$$

$$+ \frac{\rho}{2 \, \Pi X} \left[\left(\frac{\tilde{\nabla} Y_{(2)}}{X_{(2)}} + \frac{\tilde{\nabla} Y_{(1)}}{X_{(1)}} \right) \Delta X - \left(\frac{\tilde{\nabla} X_{(2)}}{X_{(2)}} + \frac{\tilde{\nabla} X_{(1)}}{X_{(1)}} \right) \Delta Y \right]^2,$$

$$(10.21)$$

where we have again used the notation $\Pi X \equiv X_{(2)} \cdot X_{(1)}$, $\Sigma X \equiv X_{(2)} + X_{(1)}$ and $\Delta X \equiv X_{(2)} - X_{(1)}$.

For two *solutions* $(X_{(A)}, Y_{(A)})$ of the field equations one has $e_{(A)} \equiv e(X_{(A)}, Y_{(A)}) = 0$ and $f_{(A)} \equiv f(X_{(A)}, Y_{(A)}) = 0$. Stokes' theorem then implies that the boundary integral over $\rho \tilde{\nabla} \{ [(\Delta X)^2 + (\Delta Y)^2] / \Pi X \}$ is equal to the sum of the integrals of the manifestly non–negative terms on the r.h.s. of eq. (10.21). Since the former vanishes as a consequence of the boundary conditions, the integrands on the r.h.s. must vanish as well.

A look at Robinson's identity (10.21) shows that it is certainly an extremely difficult, if not impossible, task to find its generalization to the electromagnetic case. Hence, it is not surprising that the earlier attempts which were based on ingenious trial and error methods failed, until eventually Bunting and Mazur came up with a systematic approach. In fact, Robinson's identity is recovered from Mazur's identity (10.1) for the vacuum case, $\Lambda = 0$: Using the parametrization (10.4) for the positive hermitian $SU(1,1)$–valued matrix Φ in terms of the Kinnersley vector v, as well as the expression (10.5) for v in terms of the Ernst potential E (and $\Lambda = 0$), one finds

$$\Phi = \frac{1}{2X} \begin{pmatrix} |E|^2 + 1 & |E|^2 + (\overline{E} - E) - 1 \\ |E|^2 + (E - \overline{E}) - 1 & |E|^2 + 1 \end{pmatrix}.$$

From this one also obtains the matrix $\Psi = \Phi_{(2)}\Phi_{(1)}^{-1} - \mathbb{1}$, which parametrizes the difference of two solutions,

$$(2\Pi X)\, \Psi =$$

$$\begin{pmatrix} (\Delta X)^2 + (\Delta Y)^2 + i\,I_- & \Delta X \Sigma X + \Delta Y \Sigma Y - i\,I_+ \\ \Delta X \Sigma X + \Delta Y \Sigma Y + i\,I_+ & (\Delta X)^2 + (\Delta Y)^2 - i\,I_- \end{pmatrix},$$

where $I_\pm \equiv Y_{(2)}|E_{(1)}^2| - Y_{(1)}|E_{(2)}|^2 \pm \Delta Y$. As is seen from the trace of this matrix, the divergence terms in Mazur's identity (10.1) and Robinson's identity (10.21) are indeed identical,

$$- \tilde{d}^\dagger (\rho\, d\, \mathrm{Tr}\, \Psi) = \tilde{\nabla} \cdot \left[\rho \tilde{\nabla} (\frac{(\Delta X)^2 + (\Delta Y)^2}{\Pi X}) \right]. \qquad (10.22)$$

Computing the second term on the l.h.s. and the term on the r.h.s. of the Mazur identity (10.1) yields, after some simple but unpleasant algebra, the corresponding terms on the r.h.s. of Robinson's identity (10.21).

For the sake of completeness, we also give the explicit expressions for the $SU(2,1)$ matrix Φ, playing the relevant role in the electrovac case. A short computation yields

$$(2\,X)\, \Phi =$$

$$\begin{pmatrix} |E|^2 + 2|\Lambda|^2 + 1 & |E|^2 + (\overline{E} - E) - 1 & 2\Lambda(\overline{E} - 1) \\ |E|^2 + (E - \overline{E}) - 1 & |E|^2 - 2|\Lambda|^2 + 1 & 2\Lambda(\overline{E} + 1) \\ 2\overline{\Lambda}(E - 1) & 2\overline{\Lambda}(E + 1) & 2|\Lambda|^2 - (E + \overline{E}) \end{pmatrix},$$

where $X = -\mathrm{Re}(E) - \Lambda\overline{\Lambda}$. In terms of $\epsilon = (1 + E)/(1 - E)$ and $\lambda = 2\Lambda/(1 - E)$ introduced in eq. (5.84), one has

$$(1 - |\epsilon|^2 - |\lambda|^2)\, \Phi =$$

$$\begin{pmatrix} 1 + |\epsilon|^2 + |\lambda|^2 & -2\,\epsilon & -2\,\lambda \\ -2\,\overline{\epsilon} & 1 + |\epsilon|^2 - |\lambda|^2 & 2\,\overline{\epsilon}\,\lambda \\ -2\,\overline{\lambda} & 2\,\epsilon\,\overline{\lambda} & 1 - |\epsilon|^2 + |\lambda|^2 \end{pmatrix}.$$

This enables one - in principle - to make Mazur's identity explicit, that is, to write down the electrovac generalization of the Robinson identity. However, this is superfluous and is not expected to yield more insight. Comparing the divergence term (see eq. (10.18)),

$$- \tilde{d}^\dagger \left(\rho \, d \operatorname{Tr} \Psi \right) =$$

$$\tilde{\nabla} \cdot \left[\rho \tilde{\nabla} \left(\frac{(\triangle X)^2 + |\triangle \Lambda|^2 (|\triangle \Lambda|^2 + 2\Sigma X) + (\triangle Y + \operatorname{Im}(\triangle \Lambda \Sigma \overline{\Lambda}))^2}{\Pi X} \right) \right]$$

with the corresponding vacuum expression (10.22), already indicates the impossibility of obtaining the resulting identity in a nonsystematic way. In conclusion, we note that the sigma model approach to the uniqueness theorem is an excellent example of how insight into the structure and the symmetries of a physical problem may greatly simplify its treatment.

10.5 Appendix: The sigma model Lagrangian

The objective of this appendix is to give the reasons for the existence of the matrix Φ providing a description of the nonlinear sigma model on the symmetric space G/H. We shall argue that Φ is hermitian and positive if $G/H = SU(p,q)/S(U(p) \times U(q))$. Our discussion is necessarily brief and we refer the reader to the second volume of Kobayashi and Nomizu (1969) and to Boothby (1975) for an introduction to symmetric spaces and to Eichenherr and Forger (1980x) for the subject treated in this section.

We have already mentioned that $SU(1,2)/S(U(1) \times U(2))$ is a symmetric space. More generally, let G be a connected Lie group, H a closed subgroup of G and σ an involutive automorphism of G. Then the triple (G, H, σ) is called a symmetric space if $G_0 \subset H \subset G_\sigma$, where G_σ denotes the closed subgroup of G containing the fixed points of σ, and G_0 is the identity component of G_σ.

Consider the canonical embedding $\hat{\sigma}$ of G/H in G,

$$\hat{\sigma} : \quad gH \mapsto g\sigma^{-1}(g) . \tag{10.23}$$

By virtue of $\hat{\sigma}$ one can assign a map $\hat{\phi}$ with target manifold G to the map ϕ with target manifold G/H,

$$\hat{\phi} : \quad \Gamma \to G , \quad \hat{\phi} = \hat{\sigma} \circ \phi . \tag{10.24}$$

Let θ denote the left–invariant Cartan form of G, $L_g^\star \theta = \theta$,

$$\theta : \quad T_g G \to T_e G , \quad \theta(X) = dL_{g^{-1}} X . \tag{10.25}$$

Using θ, it is possible to assign a bi–invariant Riemannian metric, r, to the $\operatorname{Ad}(G)$–invariant scalar product $\langle \, \cdot \, , \, \cdot \, \rangle$ of the Lie algebra

$\mathcal{G}$ of G,

$$r(X, Y) = \langle \theta(X), \theta(Y) \rangle. \tag{10.26}$$

Taking advantage of this metric, the harmonic Lagrangian $||d\hat{\phi}||^2$ for the map $\hat{\phi}$ can be written as follows:

$$
\begin{aligned}
||d\hat{\phi}||^2 &= \mathrm{Tr}\,(d\hat{\phi}\,|\,d\hat{\phi}) = \tilde{\gamma}^{ij}\, r\,(d\hat{\phi}\cdot e_i,\, d\hat{\phi}\cdot e_j) \\
&= \tilde{\gamma}^{ij}\,\langle\,\theta(d\hat{\phi}\cdot e_i),\,\theta(d\hat{\phi}\cdot e_j)\,\rangle \\
&= \tilde{\gamma}^{ij}\,\langle\,\hat{\phi}^\star\theta(e_i),\,\hat{\phi}^\star\theta(e_j)\,\rangle \equiv \langle\,\hat{\phi}^\star\theta\,|\,\hat{\phi}^\star\theta\,\rangle.
\end{aligned} \tag{10.27}
$$

The pull–back of θ with respect to the mapping $\hat{\phi}$ defines a $\mathcal{G}$–valued 1–form J on the base manifold Γ,

$$J = \hat{\phi}^\star\theta, \tag{10.28}$$

in terms of which eq. (10.27) becomes $||d\hat{\phi}||^2 = \langle\,J\,|\,J\,\rangle$. Since we are interested in the harmonic map ϕ with target manifold G/H, we look for an expression for $||d\phi||^2$ instead of $||d\hat{\phi}||^2$. Making use of the embedding (10.23), the definition (10.24) for $\hat{\phi}$ and the $\mathcal{G}$–valued 1–form $\hat{\theta} = \hat{\sigma}^\star\theta$ of G/H, it is a straightforward task to show that $||d\phi||^2 = ||d\hat{\phi}||^2$. Hence, the Lagrangian for the harmonic map $\phi: \Gamma \rightarrow G/H$ becomes

$$||d\phi||^2 = \langle\,J,\,J\,\rangle, \quad \text{where } J = \hat{\phi}^\star\theta = \phi^\star(\hat{\sigma}^\star\theta) = \phi^\star\hat{\theta}. \tag{10.29}$$

Consider now the group $G = SU(p, q) = \{g\,|\,g\eta g^\dagger = \eta\}$ and the involutive automorphism $\sigma: G \rightarrow G$,

$$\sigma(g) = \eta\, g\, \eta^{-1}, \quad \text{with } \eta = \mathrm{diag}\,(\underbrace{-1, \cdots, -1}_{p},\, \underbrace{1, \cdots, 1}_{q}). \tag{10.30}$$

The elements $a \in G$ which are invariant under σ are subject to $a = \eta a \eta^{-1}$, that is, $a \in H = S(U(p) \times U(q))$. The embedding $\hat{\sigma}$ of $SU(p, q)/S(U(p) \times U(q))$ in $SU(p, q)$ is given by eq. (10.23), which now yields

$$\hat{\sigma}(g) = g\,(\eta g \eta^{-1})^{-1} = g\eta g^{-1}\eta^{-1} = gg^\dagger. \tag{10.31}$$

For the example under consideration, $\hat{\sigma}(g)$ can therefore be represented by a positive hermitian matrix, which was denoted by Φ in this chapter. In terms of Φ, the $\mathcal{G}$–valued 1–form $J = \phi^\star(\hat{\sigma}^\star\theta)$ becomes (see eq. (10.29))

$$J = \Phi^{-1}\,d\Phi, \tag{10.32}$$

which supplies the link to the previous sections.

11

Scalar mappings

A natural generalization of the notion of ordinary scalar fields is provided by mappings between (pseudo-)Riemannian manifolds. Among these, the harmonic mappings (Fuller 1954), to which a large amount of mathematical literature has been devoted during the last thirty years, are especially prominent. (See Eells and Lemaire 1978, 1988 for comprehensive reviews and reference lists.) The developments in this field - initiated by the work of Eells and Samson (1964) on harmonic maps into manifolds with nonpositive curvature - have also led to considerable insight into the differential geometry of manifolds (see, e.g., Sacks and Uhlenbeck 1981, Schoen and Uhlenbeck 1982, 1983, Uhlenbeck 1989 and Li and Tian 1992).

Besides being interesting in their own right, it has also become obvious that harmonic maps have numerous applications in physics. In the simplest case, that is, for linear target spaces, harmonic maps reduce to harmonic functions, describing ordinary free scalar fields. However, if the target manifold has a nontrivial geometrical structure, the harmonic map equation becomes nonlinear. Important examples are the sigma models. In its original form, the nonlinear sigma model consists of a map into S^3, and was introduced in order to describe the meson fields π_1, π_2, π_3 and σ, subject to $\pi_1^2 + \pi_2^2 + \pi_3^2 + \sigma^2 = 1$ (Gell–Mann and Levy 1960; see also Duff and Isham 1977). More generally, one considers mappings into coset (symmetric) spaces G/H, where G denotes a Lie group and H is a closed subgroup of G. This class of models provides, for instance, an important tool for proving uniqueness theorems by means of integral identities (see chapter 10).

As Misner (1978) pointed out when introducing the subject to physicists in a well–known paper, part of the attractiveness of harmonic mappings also stems from their aesthetic features

- a property they share with gauge theories. Our treatment of harmonic mappings in this chapter will be rather modest: Our main goal is to introduce the basic notions in so far as they are relevant to the uniqueness theorems for self–gravitating harmonic fields (see chapter 12).

We start by introducing the connection $\overline{\nabla}$ *along* a mapping ϕ between two (pseudo–)Riemannian manifolds. We also give the expressions for the curvature and the torsion associated with $\overline{\nabla}$.

In the second section we introduce an action for ϕ. As already indicated, we choose the most natural possibility, that is, the harmonic action. As an illustration, we consider the special cases for which the variational equations describe either harmonic functions or geodesics on the target manifold.

The third section is devoted to mappings with a Skyrme action. From a mathematical point of view, the Skyrme Lagrangian represents the most natural extension of the harmonic Lagrangian. It was originally introduced for physical reasons, in order to provide nontrivial soliton solutions for the sigma model from $I\!R^3$ into S^3 (Skyrme 1961). Our presentation of the Skyrme model in this section is probably less familiar to physicists than the usual one, which is based on the Maurer–Cartan form of $SU(2)$, and which will be presented in the fourth section. However, in contrast to the latter, the mathematical approach (Manton 1987) is more general, since it covers arbitrary target manifolds. The motivation for dealing with the Skyrme model in this text is given by the fact that there *do* exist black holes with Skyrme "hair" (Luckock and Moss 1986, Droz *et al.* 1991), whereas the situation is completely different in the *harmonic* case (Heusler 1992, 1993, 1995b). The uniqueness theorem for black holes with self–gravitating *harmonic* scalar fields will be discussed in the next chapter.

The original formulation of the Skyrme model is given in the fourth section, where we also present an efficient derivation of the self–gravitating Skyrme equations for spherically symmetric configurations. As already mentioned, the latter admit black hole solutions with scalar "hair".

The last section is devoted to conformally coupled scalar fields. Using two generating techniques due to Janis *et al.* (1969) and Bekenstein (1974b), we conclude this chapter with a derivation of the Bekenstein solution for conformally coupled scalar fields.

11.1 Mappings between manifolds

In this section we introduce some concepts for mappings between manifolds, $\phi : (M, g) \to (N, G)$. In the following, these mappings will also be called *general* scalar fields. In particular, we define the connection along ϕ, $\overline{\nabla}$, and introduce the Laplacian, $\overline{\Delta}$, associated with $\overline{\nabla}$. The reader may also skip temporarily this section and read it later for a better understanding of the harmonic mapping equation (11.21). The outline follows the reasoning given by Straumann (1992). Unless otherwise specified, M and N are arbitrary (pseudo–)Riemannian manifolds with metrics g and G and dimensions $\dim(M)$ and $\dim(N)$, respectively. Their tangent spaces are TM and TN, and π denotes the projection map from TN into the target manifold, $\pi : TN \to N$.

Definition 11.1 *The mapping*

$$X : M \to TN \quad with \;\; \pi \circ X = \phi \qquad (11.1)$$

is called a vector field along ϕ; the set of all C^∞ vector fields along ϕ is denoted by $\mathcal{X}(\phi)$.

Of particular importance are the tangential vector fields in $\mathcal{X}(\phi)$. They are defined in terms of tangential mappings $T\phi : TM \to TN$ and ordinary vector fields $A' \in \mathcal{X}(M)$ as follows:

Definition 11.2 *The mappings A which can be represented in the form*

$$A = T\phi \circ A' \quad with \;\; A' \in \mathcal{X}(M) \qquad (11.2)$$

are called tangential vector fields along ϕ, $A \in \mathcal{X}(\phi)^T \subset \mathcal{X}(\phi)$.

The computation of covariant derivatives requires the concept of a connection associated with ϕ. This connection is induced by the connection ∇ of the *target* manifold as follows:

Definition 11.3 *Let ∇ be a linear connection on N, X a vector field along ϕ, $f \in \mathcal{F}(M)$ and $A = T\phi \circ A'$, where $A \in \mathcal{X}(\phi)^T$, $A' \in \mathcal{X}(M)$. Then the mapping $\overline{\nabla}$,*

$$\overline{\nabla} : \mathcal{X}(\phi)^T \times \mathcal{X}(\phi) \to \mathcal{X}(\phi) \quad with \;\; (A, X) \mapsto \overline{\nabla}_A X , \qquad (11.3)$$

is called the (induced) connection along ϕ if it has the following properties:

$$\overline{\nabla}_{A+B}X = \overline{\nabla}_A X + \overline{\nabla}_B X, \tag{11.4}$$

$$\overline{\nabla}_{fA}X = f\overline{\nabla}_A X, \tag{11.5}$$

$$\overline{\nabla}_A(X+Y) = \overline{\nabla}_A X + \overline{\nabla}_A Y, \tag{11.6}$$

$$\overline{\nabla}_A(fX) = f\cdot\overline{\nabla}_A X + (A'f)X, \tag{11.7}$$

$$\overline{\nabla}_A(Y\circ\phi) = (\nabla_A Y)\circ\phi, \tag{11.8}$$

where $Y\in\mathcal{X}(N)$ is defined by $X = Y\circ\phi$, and $((\nabla_A Y)\circ\phi)(p) = \nabla_{A(p)}Y\in T_{\phi(p)}N$.

Corollary 11.4 *The connection $\overline{\nabla}$ along ϕ is uniquely given in terms of ∇. Its representation in local coordinates $\{y^J\}$ of $\phi(U)\subset N$ is*

$$\overline{\nabla}_A X = (A'X^J)(\partial_J\circ\phi) + X^J\cdot(\nabla_A\partial_J)\circ\phi, \tag{11.9}$$

where $U\subset M$, $\partial_J = \partial/\partial y^J$, $X\in\mathcal{X}(\phi)$, $A\in\mathcal{X}(\phi)^T$ and $A'\in\mathcal{X}(M)$.

Proof In terms of local coordinates on $\phi(U)$, the vector field $X\in\mathcal{X}(\phi)$ has the representation

$$X = X^J(\partial_J\circ\phi)\quad\text{with }X^J\in\mathcal{F}(U). \tag{11.10}$$

Let $A = T\phi\circ A'$ with $A'\in\mathcal{X}(M)$. Since the components X^J are functions on $U\subset M$, we can use the property (11.7) to compute $\overline{\nabla}_A X$,

$$\overline{\nabla}_A X = X^J\overline{\nabla}_A(\partial_J\circ\phi) + (A'X^J)(\partial_J\circ\phi).$$

This establishes the expression (11.9): Since the ∂_J are vector fields on N, the property (11.8) yields $\overline{\nabla}_A(\partial_J\circ\phi) = (\nabla_A\partial_J)\circ\phi$. In order to prove the existence of the induced connection, one defines $\overline{\nabla}$ for $U\subset M$ with $\phi(U)\subset N$ according to eq. (11.9). Verifying the properties (11.4)-(11.8), the uniqueness of the local definition of $\overline{\nabla}_A X$ then implies the existence of the corresponding vector field in $\mathcal{X}(\phi)$. $\qquad\square$

It is now straightforward to define the torsion $\overline{T}$ and the curvature $\overline{R}$ *along* ϕ by using the corresponding quantities associated with the connection ∇ of N. One can then show that $\overline{T}$ and $\overline{R}$ are given in terms of $\overline{\nabla}$ as follows:

Corollary 11.5 *Let $A, B \in \mathcal{X}(\phi)^T$ and $X \in \mathcal{X}(\phi)$. Then the torsion and the curvature along ϕ are given in terms of the connection $\overline{\nabla}$ along ϕ by the formulae*

$$\overline{T}(A, B) = \overline{\nabla}_A B - \overline{\nabla}_B A - [A, B], \qquad (11.11)$$
$$\overline{R}(A, B)X = [\overline{\nabla}_A \overline{\nabla}_B, \overline{\nabla}_B \overline{\nabla}_A] X - \overline{\nabla}_{[A, B]} X. \ (11.12)$$

We leave the proof of this corollary as an exercise to the reader. As a hint, we note that for tangential fields A and B ($A = T\phi \circ A'$, $B = T\phi \circ B'$), $[A, B]$ is again a tangential field along ϕ, that is,

$$[A, B] = T\phi \circ [A', B']. \qquad (11.13)$$

In order to establish the above expressions, it is also convenient to use local coordinates, in terms of which a tangential vector field has the representation

$$A = T\phi \circ A' = A'(y^J \circ \phi) \partial_J \circ \phi. \qquad (11.14)$$

Let us now compute the coordinate expression for the connection along ϕ. Using local coordinates $\{x^\mu\}$ of $U \subset M$ and $\{y^J\}$ of $\phi(U) \subset N$ with $y^J = \phi^J(x)$, and choosing for A' a basis field on TM, $A' = \partial_\mu$, eq. (11.14) becomes

$$A = T\phi \circ \partial_\mu = \phi^J{}_{,\mu} \partial_J.$$

Together with $X = X^K(\partial_K \circ \phi)$ and eq. (11.9) we thus have

$$\overline{\nabla}_A X = \partial_\mu X^J (\partial_J \circ \phi) + X^J (\nabla_{\phi^K{}_{,\mu} \partial_K} \partial_J) \circ \phi.$$

In order to compute the second term, we recall that $\nabla_{\phi^K{}_{,\mu} \partial_K} \partial_J = \phi^K{}_{,\mu} \nabla_{\partial_K} \partial_J$, and use the Christoffel symbols $\Gamma^I{}_{KJ}$ of the target manifold, $\Gamma^I{}_{KJ} \partial_I = \nabla_{\partial_K} \partial_J$, to write

$$X^J \nabla_{\phi^K{}_{,\mu} \partial_K} \partial_J \circ \phi = \phi^K{}_{,\mu} X^J \Gamma^I{}_{KJ} \circ \phi (\partial_I \circ \phi).$$

This gives the result

$$\overline{\nabla}_{T\phi \circ \partial_\mu} X = (\partial_\mu X^I + \phi^K{}_{,\mu} \Gamma^I{}_{KJ} \circ \phi X^J) \partial_I \phi \equiv \overline{\nabla}_\mu X^I (\partial_I \circ \phi),$$

from which we finally obtain the following coordinate representation for the connection along ϕ:

$$\overline{\nabla}_\mu X^I = \partial_\mu X^I + \Gamma^I{}_{\mu J} X^J,$$
$$\text{where } \Gamma^I{}_{\mu J} \equiv \phi^K{}_{,\mu} \Gamma^I{}_{KJ} \circ \phi. \qquad (11.15)$$

We finally consider the differential $d\phi$ of the map $\phi : M \to N$. For every point $x \in M$, $d\phi(x)$ maps the tangent space $T_x M$ linearly to $T_{\phi(x)} N$. Using local coordinates on M and N, one has the representation

$$d\phi = \phi^J{}_{,\nu}(x)\,(\partial_J \circ \phi)\,dx^\nu\,, \qquad (11.16)$$

which shows that the components $\phi^J{}_{,\nu}$ of $d\phi$ are elements of the tensor bundle $\phi^*(TN) \otimes T^*M$ (see, e.g., Misner 1978). The co-variant derivative along ϕ of the $\phi^J{}_{,\nu}$ is now obtained from eq. (11.15),

$$\begin{aligned}
\overline{\nabla}_\mu \phi^J{}_{,\nu} &= \nabla_\mu \phi^J{}_{,\nu} + \Gamma^J{}_{\mu K}\,\phi^K{}_{,\nu} \\
&= \phi^J{}_{,\nu}{}_{,\mu} - \Gamma^\sigma{}_{\mu\nu}\,\phi^J{}_{,\sigma} + \Gamma^J{}_{IK}\,\phi^I{}_{,\mu}\,\phi^K{}_{,\nu}\,, \quad (11.17)
\end{aligned}$$

where we have written $\Gamma^J{}_{IK}$ for $\Gamma^J{}_{IK} \circ \phi$. The above formula also suggests the following definition of the Laplacian along ϕ:

$$\overline{\Delta}\phi^J = g^{\mu\nu}\overline{\nabla}_\mu \phi^J{}_{,\nu} = \Delta\phi^J + \Gamma^J{}_{IK}\,(d\phi^I | d\phi^K)\,, \qquad (11.18)$$

where Δ and $(\cdot|\cdot)$ denote the Laplacian and the inner product with respect to g. As will be discussed in the next section, $\overline{\Delta}$ is the relevant operator associated to variations of the harmonic density $G_{IJ}(d\phi^I | d\phi^J)$. It is also known as the tension field and is denoted by τ: $\overline{\Delta}\phi^J \equiv \tau^J(\phi)$ (see, e.g., Eells and Lemaire 1978).

11.2 Harmonic mappings

As already mentioned, harmonic mappings play a distinguished role amongst the mappings between (pseudo–)Riemannian manifolds. In fact, they provide the natural extension of harmonic functions to nonlinear target spaces. The harmonic action is the pull–back of the line element of the target manifold. More precisely, we have the following definition:

Definition 11.6 *Let $\langle \cdot, \cdot \rangle$ denote the inner product with respect to both the metric g of the base manifold M and the metric G of the target manifold N, and let $\eta = *1$ be the volume–form on (M, g). The quantities $e_2[\phi]$ and $S_2[\phi]$,*

$$e_2[\phi] \equiv \frac{1}{2}\langle d\phi, d\phi \rangle\,, \quad S_2[\phi] \equiv \int_M e_2[\phi]\,\eta\,, \qquad (11.19)$$

are called the harmonic energy density and the harmonic action of ϕ, respectively.

Note that in local coordinates on M and N, the harmonic energy density has the representation

$$e_2[\phi] = \frac{1}{2} G_{AB}(\phi) \, (d\phi^A | d\phi^B) = \frac{1}{2} G_{AB}(\phi(x)) \, g^{\mu\nu}(x) \, \phi^A{}_{,\mu} \, \phi^B{}_{,\nu} \, .$$
(11.20)

We now compute the first variation of $S_2[\phi]$. (As usual, we assume that the quantities involved in the calculation, such as the vector field generating the variation, are sufficiently differentiable (see, e.g., Hungerbühler 1994). More precisely, we consider the class F of mappings $M \to N$ defined by certain boundary conditions on ∂M (if there is a boundary) and assume that the solution of the variational problem is an element of F. For small values of $|t|$, the variations $\phi_t(x) : M \to N$ around $\phi = \phi_0$ are required to coincide with ϕ_0 outside a compact set in M and to be C^1 in x and t. Assuming further that the vector field $v(x) = [\frac{d}{dt}\phi_t(x)]_{t=0} \in T_{\phi(x)}N$ is also C^1, one knows that $e_2[\phi_t]$ is finite, which enables one to bring the differentiation with respect to t under the integral. Provided that ϕ is at least C^2, one may finally integrate by parts.) The variation of the harmonic density becomes

$$
\begin{aligned}
\delta e_2 \eta &= d\delta\phi^A \wedge G_{AB} * d\phi^B + \frac{1}{2} \delta G_{AB} \, d\phi^A \wedge *d\phi^B \\
&\doteq -\delta\phi^C \wedge d\left[G_{CB} * d\phi^B\right] + \frac{1}{2} \delta\phi^C \, G_{AB,C} \, d\phi^A \wedge *d\phi^B \\
&= \delta\phi^C \left[\left(\frac{G_{AB,C}}{2} - G_{CB,A}\right) d\phi^A \wedge *d\phi^B + G_{CB} \, d * d\phi^B\right] \\
&= -\delta\phi^C G_{CD} \left[\Gamma^D_{AB}(d\phi^A | d\phi^B) + \Delta\phi^D\right] \eta,
\end{aligned}
$$

where, as usual, '$\doteq$' stands for equal up to a total derivative, here $d(G_{AB}\delta\phi^A * d\phi^B)$. From this, one concludes that the variation of the harmonic action (11.19) vanishes if the *nonlinear* Euler–Lagrange equations,

$$\overline{\Delta}\phi^A = \Delta\phi^A + \Gamma^A_{BC}(d\phi^B | d\phi^C) = 0, \qquad (11.21)$$

are satisfied, that is, if ϕ is a harmonic mapping with respect to the Laplacian along ϕ (see eq. (11.18)). To illustrate this equation, we consider the following simple examples:

(i) Let N be the 1–dimensional target manifold $\mathbb{R}$. Then the mapping $\phi : M \to \mathbb{R}$ is a real function and the harmonic density becomes $e_2[\phi] = \frac{1}{2}(d\phi|d\phi)$. The Laplacian along ϕ reduces to the ordinary Laplacian Δ with respect to the metric $\boldsymbol{g}$ of M. Hence, the Euler–Lagrange equation (11.21) then becomes the usual linear equation $\Delta\phi = 0$, characterizing *harmonic functions* on M.

(ii) Let N be a vector space. Then the connection on the target manifold is trivial, $\Gamma^C_{AB} = 0$. The situation is therefore the same as above, where now $\Delta\phi^A = 0$ are $n \equiv \dim(N)$ linear, decoupled equations.

(iii) Let the base manifold be 1–dimensional, for instance $M = I$ $\in \mathbb{R}$. Then $\phi : I \to N$, $t \in [t_0, t_1] \mapsto \phi(t)$ describes a curve on the target space $(N, \boldsymbol{G})$ with $e_2[\phi] = \frac{1}{2}G_{AB}(\phi)\partial_t\phi^A\partial_t\phi^B$. The Laplacian on the base manifold reduces to the ordinary second derivative with respect to t, and the Euler–Lagrange equation (11.21) becomes the geodesic equation on the target manifold:

$$\frac{d^2\phi^A}{dt^2} + \Gamma^A_{BC}(\phi)\frac{d\phi^B}{dt}\frac{d\phi^C}{dt} = 0. \qquad (11.22)$$

(iv) The nonlinear sigma model: Consider $n + 1$ real functions φ^a, subject to $\sum^{n+1}\varphi^a\varphi^a \equiv \vec{\varphi}\cdot\vec{\varphi} = 1$. Since the φ^a parametrize the n–sphere in $\mathbb{R}^{n+1}$, one can equivalently consider the map $\phi : M \to S^n$, where now the $n \equiv \dim(N)$ fields ϕ^A are independent. The variation of the sigma model action with constraint,

$$S_\lambda[\varphi] = \int \left[\frac{1}{2}(d\vec{\varphi}|d\vec{\varphi}) - \lambda(\vec{\varphi}\cdot\vec{\varphi} - 1)\right]\eta,$$

with respect to φ^a and λ yields the equation

$$\Delta\vec{\varphi} + (d\vec{\varphi}|d\vec{\varphi})\,\vec{\varphi} = 0. \qquad (11.23)$$

(Use $2\lambda = -\vec{\varphi}\Delta\vec{\varphi} = d^\dagger(\vec{\varphi}d\vec{\varphi}) + (d\vec{\varphi}|d\vec{\varphi})$ and $\vec{\varphi}d\vec{\varphi} = 0$ to derive this.) This is, in fact, equivalent to the general equation (11.21) for the ϕ^A, where Γ^A_{BC} are the Christoffel symbols of S^n. As an exercise, we suggest the reader considers the parametrization $\vec{\varphi} = (\sin\phi^1\cos\phi^2, \sin\phi^1\sin\phi^2, \cos\phi^1)$ of the 2–sphere in $\mathbb{R}^3$ with metric $\boldsymbol{G}(\phi) = d(\phi^1)^2 + \sin^2\phi^1 d(\phi^2)^2$ and verifies the equivalence of eqs. (11.21) and (11.23) explicitly.

(v) In view of later applications, we also consider harmonic maps from a spherically symmetric spacetime M into the 3–sphere S^3 with metric $\boldsymbol{G} = d(\phi^1)^2 + \sin^2\phi^1(d(\phi^2)^2 + \sin^2\phi^2 d(\phi^3)^2)$. Denoting the coordinates on M by r, t, ϑ and φ, the spacetime metric is $^{(4)}\boldsymbol{g} = \tilde{\boldsymbol{g}}(r, t) + r^2 d\Omega^2$, where $\tilde{\boldsymbol{g}}$ denotes a 2–dimensional pseudo–Riemannian metric. Restricting our attention to spherically symmetric mappings,

$$\phi^1 = \chi(r,t), \quad \phi^2 = \vartheta, \quad \phi^3 = \varphi, \tag{11.24}$$

the harmonic action becomes

$$S_2[\phi] = 4\pi \int e_2[\chi] \, r^2 \tilde{\eta}, \quad e_2[\chi] = \frac{1}{2}(d\chi|d\chi) + \frac{\sin^2\chi}{r^2}, \tag{11.25}$$

where $\tilde{\eta}$ and $(\cdot|\cdot)$ refer to the metric $\tilde{g}$.

We conclude this section by giving a Bochner (1940) identity, first derived by Eells and Samson (1964) for the harmonic density $e_2[\phi]$. The identity will be used in section 12.4 to establish uniqueness theorems for a class of harmonic mappings. (As we shall demonstrate in section 12.5, there exist, however, different methods which are more powerful, since they cover a larger class of harmonic maps.) The Bochner identity is obtained as follows: Using the property

$$\overline{\nabla}_\mu \phi^A{}_{,\nu} = \overline{\nabla}_\nu \phi^A{}_{,\mu}$$

(see eq. (11.17)) and the Ricci identity (1.4), one finds after some algebraic manipulations

$$\begin{aligned}
\Delta \, e_2[\phi] &= \phi^{A:\mu}(\overline{\Delta}\phi_A)_{:\mu} + \phi^{A:\mu:\nu}\phi_{A:\mu:\nu} + \phi^A{}_{,\mu}\,\phi^B{}_{,\nu} \times \\
&\quad \left[G_{AB}R^{\mu\nu} - R_{ABCD}g^{\mu\alpha}g^{\nu\beta}\phi^C{}_{,\alpha}\phi^D{}_{,\beta} \right],
\end{aligned} \tag{11.26}$$

where the colon and $\overline{\Delta}$ denote the covariant derivative and the Laplacian along ϕ, respectively (see eqs. (11.17), (11.18)). Note that in the trivial case, $N = \mathbb{R}$, the above formula reduces to the identity

$$\Delta \frac{1}{2}(d\phi|d\phi) = \phi^\mu(\Delta\phi)_{;\mu} + \phi_{\mu;\nu}\phi^{\mu;\nu} + R^{\mu\nu}\phi_{,\mu}\,\phi_{,\nu}, \tag{11.27}$$

which is immediately verified from the ordinary Ricci identity, $\phi^\nu_{;\mu;\nu} = (\Delta\phi)_{;\mu} + R_{\mu\nu}\phi^\nu$.

11.3 Skyrme mappings

As we shall argue in the next chapter, there exist a number of uniqueness theorems for harmonic mappings - with and without taking gravitational interactions into account. This motivates the consideration of mappings with more general Lagrangians.

The original *physical* reason for extending the harmonic action was based on the expectation that in the low–energy limit of QCD, bosons might emerge as soliton solutions of an effective nonlinear sigma model Lagrangian (see Witten 1983). However, as is seen from a scaling argument, such (flat space) solitons can not exist for the ordinary sigma model (see the previous section, example (iv)). In order to overcome this problem one can, for instance, take higher order terms into account. The most natural realization of this idea is to consider an $SU(2) \times SU(2)$ invariant Lagrangian, first introduced by Skyrme (1961). A brief description of the Skyrme model in its usual form is given in the next section.

From the point of view which was assumed in the preceding sections, the Skyrme term arises in a very canonical way: The fact that the harmonic density $e_2[\phi]$ is the pull–back of the line element of the target manifold (N, G), suggests the introduction of an additional term $e_4[\phi]$, having the geometrical interpretation of the area element of (N, G) (see Esteban 1986, Manton and Ruback 1986, Manton 1987). (A third contribution, corresponding to the volume element, can be considered as well. In fact, there exists a physical motivation for doing so: The Skyrme Lagrangian describes the π and σ fields, whereas the volume term might be responsible for effects which are related to the ω mesons (Jackson *et al.* 1985). Here we restrict ourselves to the Skyrme term.)

Definition 11.7 *Let* $\phi : (M, g) \to (N, G)$. *The quantities* $e_S[\phi]$ *and* $S_S[\phi]$,

$$e_S[\phi] \equiv e_2[\phi] + e_4[\phi], \qquad S_S[\phi] \equiv \int_M e_S[\phi]\,\eta \qquad (11.28)$$

are the Skyrme energy density and the Skyrme action assigned to ϕ, *respectively. As before,* $e_2[\phi]$ *is the harmonic density (11.19), and*

$$
\begin{aligned}
e_4[\phi] &\equiv \frac{1}{4}\langle d\phi \wedge d\phi,\, d\phi \wedge d\phi \rangle \\
&= \frac{1}{4} G_{AC} G_{BD}(d\phi^A \wedge d\phi^B \,|\, d\phi^C \wedge d\phi^D).
\end{aligned}
\qquad (11.29)
$$

As an example, we evaluate this expression for the case where the base manifold is a spherically symmetric spacetime, the target manifold is the 3–sphere and ϕ is spherically symmetric in the sense that it maps 2–spheres into 2–spheres (see section 11.2, example (v)). Writing $e_4[\phi]$ in the form

$$\frac{1}{4} G_{AC} G_{BD} \left[(d\phi^A|d\phi^C)(d\phi^B|d\phi^D) - (d\phi^A|d\phi^D)(d\phi^B|d\phi^C) \right] ,$$

a short computation gives

$$e_4[\phi] = \frac{\sin^2 \chi}{r^2} \left[(d\chi|d\chi) + \frac{\sin^2 \chi}{2\,r^2} \right]. \tag{11.30}$$

Denoting the volume–form of the 2–dimensional metric $\tilde{g}$ by $\tilde{\eta}$, and recalling that $^{(4)}g = \tilde{g} + r^2 d\Omega^2$, the spherically symmetric Skyrme action (11.28) becomes

$$\begin{aligned}
S_S[\phi] = \;& 4\pi \int r^2 \tilde{\eta} \left[\left(\frac{(d\chi|d\chi)}{2} + \frac{\sin^2 \chi}{r^2} \right) \right. \\
& \left. + \frac{\sin^2 \chi}{r^2} \left((d\chi|d\chi) + \frac{\sin^2 \chi}{2\,r^2} \right) \right]. \tag{11.31}
\end{aligned}$$

Topological bounds for the energy of *static* solutions exist in various classical field theories (Bogomol'nyi 1976, Coleman *et al.* 1977). In its usual formulation, the static (and flat) Skyrme model describes mappings U from the compactified $\mathbb{R}^3$ into $SU(2)$. The Bogomol'nyi bound, which involves the degree $\Pi_3(S^3) = Z$ of U, is then derived directly from the action integral (see the next section). For the moment we keep the framework somewhat more general and, following Manton (1987), prove the following theorem:

Theorem 11.8 *Let $\phi : M \to N$ be a solution of the variational equations for the Skyrme action (11.28). Let M and N be orientable, connected Riemannian manifolds, where $\dim(M) = 3$ and N has finite volume. Then the total energy $S_S[\phi]$ may be estimated in terms of the degree of ϕ by*

$$S_S[\phi] \geq 3 \deg(\phi) \operatorname{vol}(N). \tag{11.32}$$

Note that this estimate holds without further restrictions on the geometry of either of the two manifolds. In addition, it does not depend on the dimension of the target space. The *topological* requirements are, of course, essential.

Proof Consider the orthonormal frames $\{e^\mu_\alpha\}$ and $\{E^i_A\}$ on M and N, respectively, $e^\mu_\alpha e^\nu_\beta g_{\mu\nu} = \delta_{\alpha\beta}$, $E^i_A E^j_B G^{AB} = \delta^{ij}$. Define the $3 \times \dim(N)$ matrices $\boldsymbol{A}$ according to

$$A_{\alpha i} = E^i_A \, \phi^A{}_{,\mu} \, e^\mu_\alpha \,.$$

In terms of $\boldsymbol{A}$, the energy densities $e_2[\phi]$ and $e_4[\phi]$ (see eqs. (11.20) and (11.29)) become

$$e_2 = \frac{1}{2} \operatorname{tr}(\boldsymbol{A} \cdot \boldsymbol{A}^T), \tag{11.33}$$

$$e_4 = \frac{1}{4} \left[\operatorname{tr}^2(\boldsymbol{A} \cdot \boldsymbol{A}^T) - \operatorname{tr}(\boldsymbol{A} \cdot \boldsymbol{A}^T)^2 \right]. \tag{11.34}$$

Since $\boldsymbol{A} \cdot \boldsymbol{A}^T$ is a positive definite 3×3 matrix, its eigenvalues λ_1^2, λ_2^2 and λ_3^2 are positive and the invariants e_2 and e_4 are equal to $\frac{1}{2} \sum \lambda_i^2$ and $\frac{1}{2} \sum_{i<j} \lambda_i^2 \lambda_j^2$, respectively. Hence, we find $e_2 + e_4 = \frac{1}{2}(\lambda_1 - \lambda_2\lambda_3)^2 + \lambda_1\lambda_2\lambda_3 + \text{cycl.} \geq 3\lambda_1\lambda_2\lambda_3$. This yields the inequality

$$e_2 + e_4 \geq 3 \det{}^{1/2}(\boldsymbol{A} \cdot \boldsymbol{A}^T) = 3 \det(\boldsymbol{A}).$$

Noting that $\phi^A{}_{,\mu}$ is the Jacobian of the map ϕ and that $\det(\boldsymbol{A}) = \sqrt{G} \det(\phi) \sqrt{g}^{-1}$, yields the desired result

$$S_S[\phi] = \int_M (e_2 + e_4)\, \eta \geq 3 \deg(\phi)\operatorname{vol}(N) \,,$$

where we have also used the fact that the manifolds are orientable. □

As an illustration, we consider spherically symmetric Skyrme mappings into S^3 (see above and section 11.2, example (v)). In order to have a 3–dimensional Riemannian base manifold, we require that the spacetime is *static* with metric ${}^{(4)}g = \tilde{g} + r^2 d\Omega^2$, where $\tilde{g} = -dt^2 + f^{-1}(r)dr^2$. After a formal time integration, the action (11.31) becomes

$$S_S^{st}[\phi] = 4\pi \int_0^R \left[\frac{f\chi'^2}{2}(r^2 + 2s^2) + \frac{s^2}{2r^2}(2r^2 + s^2) \right] \frac{dr}{\sqrt{f}}, \tag{11.35}$$

where $s \equiv \sin\chi$. In order for ϕ to have finite energy, we must impose the boundary conditions $\chi(0) = 0$ (without loss of generality) and $\chi(R) = n\pi$. Here n must be an integer and the value of R depends on the geometry of the static 3–surface. (If the latter is a geometric 3–sphere then $R = 1$ and $N(r) = 1 - r^2$. If, in contrast, spacetime is asymptotically flat, then the boundary condition at $R = \infty$ still enables one to compactify the spacelike hypersurface.) However, since the derivation of the bound (11.32) depends only on the topology of the manifolds, both the numerical value of R and the explicit form of the metric function $f(r)$ are irrelevant in this context. In fact, the above action can be written as

$$S_S^{st}[\phi] \;=\; 4\pi \int_0^R 3\chi' \sqrt{f} \sin^2 \chi \, \frac{dr}{\sqrt{f}}$$

$$+ 4\pi \int_0^R \frac{[(r^2 + 2s^2)\sqrt{f}\chi' - 3s^2]^2 + 2r^{-2}s^2(r^2 - s^2)^2}{2\,(r^2 + 2s^2)} \, \frac{dr}{\sqrt{f}} \, .$$

Observing that the second integrand is non–negative, and changing the integration variable from r to χ, gives the desired result,

$$S_S^{st}[\phi] \;\geq\; 3 \cdot 4\pi \int_0^{n\pi} \sin^2 \chi \, d\chi \;=\; 3\,n \, \mathrm{vol}(S^3)\,,$$

where n is identified with the degree of the map. (If the timelike Killing field does not have unit norm, then the bound is slightly modified. Here we have chosen $-V = 1$ in the static metric in order to obtain a mapping with 3–dimensional base manifold, as is required in the previous theorem.) The same bound is obtained in the next section for the Skyrme model in the "usual" setting.

It is also interesting to search for solutions which *saturate* the topological bound. Clearly, such configurations are stable due to topological reasons. In the spherically symmetric example above, the bound is attained if and only if

$$r = \sin \chi \quad \text{and} \quad \frac{d\chi}{dr} = f^{-1/2}.$$

Hence, $f^{-1}(dr|dr) = (d\chi|d\chi)$, and the metric of the spacelike hypersurface coincides with the metric G of the target manifold S^3. We therefore conclude that the Bogomol'nyi bound for spherically symmetric Skyrme maps into S^3 is attained if and only if the base manifold is also a geometrical 3–sphere, and the mapping is the identity. In fact, this result remains true even without imposing spherical symmetry on the base space: In the next section we shall see that the only Skyrme mapping from a 3–dimensional Riemannian manifold into $SU(2)$ attaining the bound (11.32) is the identity from S^3 into $SU(2) \sim S^3$ (Manton 1987).

11.4 The SU(2) Skyrme model

As already mentioned, Skyrme (1961) introduced the additional term to the quadratic Lagrangian in order to describe solitons in the low–energy limit of QCD (see also Skyrme 1962, 1971, Pak and Tze 1979, Adkins *et al.* 1983, Witten 1983 and Zahed and Brown 1986). The $SU(2) \times SU(2)$ invariant Skyrme Lagrangian density is

$$\mathcal{L}_S = \frac{f^2}{4} \mathrm{Tr}\{\nabla_\mu U \nabla^\mu U^\dagger\} - \frac{1}{32e^2} \mathrm{Tr}\{ [\nabla_\mu UU^\dagger, \nabla_\nu UU^\dagger]^2 \}, \quad (11.36)$$

where ∇ denotes the covariant derivative with respect to the spacetime metric $^{(4)}g$, e is the dimensionless (rho meson / pionic current) coupling constant, f ($\simeq 190\,\mathrm{MeV}$) is the pion decay constant and $U(x)$ is the $SU(2)$–valued function describing the meson fields. It is very convenient to rewrite the Skyrme Lagrangian in terms of the antihermitian $SU(2)$–valued 1–form A, defined as the pull–back of the Maurer–Cartan form of $SU(2)$, $A = U_*\Theta$. By virtue of the Maurer–Cartan equation, A satisfies the identity

$$dA + A \wedge A = 0. \quad (11.37)$$

Using

$$A = U^\dagger dU, \qquad F = A \wedge A, \quad (11.38)$$

the Skyrme Lagrangian assumes the simple form

$$\mathcal{L}_S = -\frac{f^2}{4} \mathrm{Tr}(A|A) - \frac{1}{16e^2} \mathrm{Tr}(F|F). \quad (11.39)$$

The variation with respect to the fundamental field U yields

$$\frac{1}{2} \delta \mathrm{Tr} (A \wedge *A) \doteq - \mathrm{Tr} (U^\dagger \delta U \, d * A),$$

$$\frac{1}{2} \delta \mathrm{Tr} (F \wedge *F) \doteq + \mathrm{Tr} (U^\dagger \delta U \, [A, d * F]),$$

where '$\doteq$' stands for equal up to a total derivative. In order to derive the above formulae, one uses $\delta A = U^\dagger d(\delta U) - U^\dagger (\delta U)A$ and $[\alpha, \beta] = \alpha \wedge \beta - (-1)^{pq}\beta \wedge \alpha$ (for arbitrary $SU(2)$–valued p– and q–forms α, β). The general $SU(2)$ Skyrme equations for an arbitrary spacetime metric $^{(4)}g$ now become

$$d * A - \frac{1}{\mu^2} [A, d * F] = 0, \quad (11.40)$$

where $\mu = 2fe$. Since the Skyrme Lagrangian (11.39) does not depend on derivatives of the metric, the stress–energy tensor is $T = 2\partial\mathcal{L}/\partial^{(4)}g^{-1} - {}^{(4)}g\,\mathcal{L}$. Hence, Einstein's equations become

$$
\begin{aligned}
G_{\mu\nu} = \ & -\kappa\,\mathrm{Tr}\left[\left(A_\mu A_\nu - \frac{{}^{(4)}g_{\mu\nu}}{2}(A|A)\right)\right.\\
& \left. + \frac{1}{(2fg)^2}\left(F_{\mu\rho}F_\nu{}^\rho - \frac{{}^{(4)}g_{\mu\nu}}{2}(F|F)\right)\right], \quad (11.41)
\end{aligned}
$$

where $\kappa = 4\pi(f/m_{Pl})^2$.

Let us now consider static, finite energy solutions on a fixed, asymptotically flat, static spacetime $M = \mathbb{R} \times \Sigma$ with metric ${}^{(4)}g = -dt^2 + g$. By requiring that asymptotically $U \to \mathbb{1}$, the spacelike hypersurface Σ can be compactified such that U describes a mapping from a topological 3–sphere into $SU(2) \sim S^3$. Hence, the static Skyrme solutions fall into homotopy classes, $\Pi_3(SU(2)) = Z$. In agreement with the general theorem 11.8, one finds that the energy of such soliton solutions is bounded from below by their topological charge:

Theorem 11.9 *Consider a static spacetime $(M, {}^{(4)}g)$ with metric ${}^{(4)}g = -dt^2 + g$ and connected, asymptotically flat spacelike hypersurface (Σ, g). Let $U : \Sigma \to SU(2)$ be a static, finite energy solution with Skyrme Lagrangian (11.39). Then the Skyrme energy is bounded from below by*

$$
E[U] \geq 3 \deg(U)\,(f/e)\,2\pi^2 . \qquad (11.42)
$$

Proof Using $*A = -dt\wedge \overset{g}{*} A$ and $*F = dt\wedge \overset{g}{*} F$, we obtain the following expression from the Lagrangian (11.39) for the energy of a static configuration:

$$
E[U] = -\frac{1}{16e^2}\int_\Sigma \mathrm{Tr}\{\mu^2 A\wedge \overset{g}{*} A + F\wedge \overset{g}{*} F\}. \qquad (11.43)
$$

Denoting the scalar product with respect to the 3–metric g with $(\cdot|\cdot)$, this yields

$$
E[U] = -\frac{1}{16e^2}\int_\Sigma \mathrm{Tr}\{([\mu A + \overset{g}{*} F]\,|\,[\mu A + \overset{g}{*} F]) - 2\mu(A|\overset{g}{*} F)\}\eta
$$

$$
\geq \frac{f}{4e}\int_\Sigma \mathrm{Tr}\{A\wedge \overset{gg}{**} F\} = \frac{f}{4e}\int_\Sigma \mathrm{Tr}\{A\wedge A\wedge A\}, \qquad (11.44)
$$

where we have used the fact that $(\alpha|\alpha) = \alpha_i\alpha_j g^{ij} \leq 0$ for an antihermitian 1–form α and a Riemannian metric $\boldsymbol{g}$. In order to conclude the proof, we recall that the normalized volume–form on $SU(2)$ is (see below)

$$\eta_{SU(2)} = \frac{1}{24\pi^2} \, \mathrm{Tr} \left\{ U^\dagger dU \wedge U^\dagger dU \wedge U^\dagger dU \right\}. \tag{11.45}$$

Hence, the last integrand in eq. (11.44) is the pull–back of $\eta_{SU(2)}$ times the volume of $SU(2)$, implying that $E[U] \geq \frac{f}{4e} 24\pi^2 \deg(U)$.

$\square$

(Note that a factor of $\sqrt{V}$ appears in the integrals in eq. (11.44) if the Killing field is not unit normalized, i.e., if $^{(4)}\boldsymbol{g} = -Vdt^2 + \boldsymbol{g}$.)

Let us eventually derive the full set of field equations for the self–gravitating, spherically symmetric case. The most efficient way to do this is by expressing the Maurer–Cartan form of $SU(2)$ in terms of a conveniently chosen basis of hermitian matrices and 1–forms on S^3:

Proposition 11.10 *Let $\{\theta^i\}$ be the orthonormal standard basis of 1–forms on S^3 with respect to spherical coordinates χ, ϑ and φ. Then there exist matrices $\{\hat\sigma_i\}$, with $[\hat\sigma_i, \hat\sigma_j] = 2i\epsilon_{ijk}\hat\sigma_k$ and $\hat\sigma_i^2 = \mathbb{1}$, such that the Maurer–Cartan form of $SU(2)$ is given by*

$$\Theta = i\left[\hat\sigma_1\theta^1 + \hat\sigma_2\theta^2 + \hat\sigma_3\theta^3\right]. \tag{11.46}$$

In addition, Θ fulfils the identity

$$\Theta \wedge \Theta + 2 \, \dot\ast \, \Theta = 0, \tag{11.47}$$

where $\dot\ast$ denotes the Hodge dual on S^3, $\dot\ast\theta^1 = \theta^2 \wedge \theta^3$ (and cyclic).

Proof Let $\{y^i\}$ be standard coordinates on $\mathbb{R}^4$ and consider the embedding $(y^4)^2 + |\underline{y}|^2 = 1$ of S^3 in $\mathbb{R}^4$. Then

$$U = y^4 \mathbb{1} + i\underline{y} \cdot \underline\sigma = \mathbb{1} \cos\chi + i\sigma_R \sin\chi. \tag{11.48}$$

Here $\underline\sigma$ denote the ordinary Pauli matrices and the third angle on S^3 is defined by $\cos\chi = y^4$, $\sin\chi = |\underline{y}|$. In addition, we have introduced the matrix $\sigma_R = \hat{R} \cdot \underline\sigma$, where $\hat{R}$ is the radial unit vector in $\mathbb{R}^3$, $\hat{R} = \underline{y}/|\underline{y}|$. Also introducing the orthonormal basis vectors

$\hat{\vartheta}$ and $\hat{\varphi}$, we have $d\hat{R} = \hat{\vartheta}d\vartheta + \hat{\varphi}\sin\vartheta d\varphi$, and thus $d\sigma_R = \sigma_\vartheta d\vartheta + \sigma_\varphi \sin\vartheta d\varphi$. Now using the orthonormal tetrad on S^3, $\theta^1 = d\chi$, $\theta^2 = \sin\chi d\vartheta$, $\theta^3 = \sin\chi \sin\vartheta d\varphi$, we find

$$dU = (-\sin\chi + i\sigma_R \cos\chi)\,\theta^1 + i\,(\sigma_\vartheta\,\theta^2 + \sigma_\varphi\,\theta^3). \qquad (11.49)$$

Eventually taking advantage of the fact that $\sigma_R^2 = \mathbb{1}$, $[\sigma_\vartheta, \sigma_\varphi] = 2i\sigma_R$ (and cyclic), we obtain the desired result (11.46) for $U^\dagger dU$, where

$$\hat{\sigma}_1 \equiv \sigma_R, \qquad \hat{\sigma}_{2,3} \equiv \cos\chi\,\sigma_{\vartheta,\varphi} \pm \sin\chi\,\sigma_{\varphi,\vartheta}. \qquad (11.50)$$

Equation (11.47) is a consequence of the fact that the Hodge dual and the commutator have a similar effect for $SU(2)$:

$$\begin{aligned}
\Theta \wedge \Theta &= -\sum_{i<j} [\hat{\sigma}_i, \hat{\sigma}_j]\,\theta^i \wedge \theta^j = -2i \sum_{i<j,k} \hat{\sigma}_k\,\epsilon_{kij}\,\theta^i \wedge \theta^j \\
&= -2i \sum_k \hat{\sigma}_k * \theta^k = -2 * \Theta.
\end{aligned}$$

$\square$

Before we proceed, we consider the following two applications of the above proposition:

(i) The explicit expression (11.46) for the Maurer–Cartan form enables a verification of the normalization factor of the volume–form (11.45) of $SU(2)$: Using eq. (11.47) one has

$$\begin{aligned}
\mathrm{Tr}\{\Theta \wedge \Theta \wedge \Theta\} &= -2\,\mathrm{Tr}\{\Theta \wedge *\Theta\} = -2\,\mathrm{Tr}\,(\Theta|\Theta)\dot{\eta} \\
&= -2\sum_j \mathrm{Tr}\{i\hat{\sigma}_j\}^2\,(\theta^j|\theta^j)\,\dot{\eta} = 12\,\dot{\eta},
\end{aligned}$$

where $\dot{\eta}$ denotes the volume form on S^3, $\dot{\eta} = \sin^2\chi d\chi \wedge d\Omega$. We thus obtain the desired result,

$$\int_{SU(2)} \eta_{SU(2)} = \frac{12}{24\pi^2}\int_{S^3}\dot{\eta} = \frac{1}{2\pi^2}\,4\pi \int_0^\pi \sin^2\chi d\chi = 1.$$

(ii) As we have demonstrated in the previous section, the topological bound (11.42) is saturated only if the base manifold is a geometric 3–sphere. In the present context, this result is recovered as follows: First, eq. (11.44) implies that $\mu A + \overset{g}{*} F$ must vanish. Taking advantage of eq. (11.47), this yields

$$\mu A + \overset{g}{*} F = \mu\Theta + \overset{g}{*}(\Theta \wedge \Theta) = \mu\Theta - 2\overset{g}{*}*\Theta = 0,$$

(where we have not written out the pull–backs). Recall that $\overset{g}{*}$ and $*$ denote the Hodge duals with respect to the 3–dimensional metrics of the base manifold Σ and the target manifold S^3, respectively. Hence, up to a constant factor, the metrics must be identical, that is $\overset{g}{*} *\Theta = (fe)\,\Theta$.

We are now prepared to derive the full set of spherically symmetric equations. The "hedgehog" Lagrangian is obtained as follows: Using the metric $^{(4)}g = \tilde{g}(r,t) + r^2 d\Omega^2$ and the ansatz $\chi = \chi(r,t)$, the expression (11.46) for the Maurer–Cartan form yields

$$-\frac{1}{4}\mathrm{Tr}\,(A|A) = \frac{1}{2}[(d\chi|d\chi) + (\theta^2|\theta^2) + (\theta^3|\theta^3)]$$

$$= \frac{1}{2}(d\chi|d\chi) + \frac{\sin^2\chi}{r^2} = e_2[\chi], \qquad (11.51)$$

$$-\frac{1}{16}\mathrm{Tr}\,(F|F) = \frac{1}{2}[(d\chi|d\chi)\,((\theta^2|\theta^2) + (\theta^3|\theta^3)) + (\theta^2|\theta^2)(\theta^3|\theta^3)]$$

$$= \frac{\sin^2\chi}{r^2}\left[(d\chi|d\chi) + \frac{\sin^2\chi}{2r^2}\right] = e_4[\chi]. \qquad (11.52)$$

Here we have used the fact that θ^2 and θ^3 are normalized with respect to the metric on S^3, whereas $(\cdot|\cdot)$ is the scalar product with respect to the spacetime metric. Thus, $(\theta^2|\theta^2) = \sin^2\chi(d\vartheta|d\vartheta) = r^{-2}\sin^2\chi = (\theta^3|\theta^3)$. (The above expressions coincide, of course, with the corresponding results (11.25) and (11.30), derived in the previous section for spherically symmetric Skyrme maps with target manifold S^3.) The Lagrangian now becomes $\mathcal{L}_S = f^2 e_2[\chi] + e^{-2}e_4[\chi]$. In order to get rid of the coupling constants, we consider the conformal transformation $\tilde{g}_{ij} \rightarrow (fe)^{-2}\tilde{g}_{ij}$ and replace r by $(fe)^{-1}r$ and $\mathcal{L}_S$ by $(fe)^2\mathcal{L}_S$. Integrating the action over the angular variables, this yields the *effective* Lagrangian, $L_S = 4\pi r^2\sqrt{\tilde{g}}\mathcal{L}_S$:

$$L_S = 2\pi f^2\,\sqrt{\tilde{g}}\,[(d\chi|d\chi)\,u(\chi) + v(\chi)], \qquad (11.53)$$

where

$$u(\chi) = r^2 + 2\sin^2\chi, \qquad v(\chi) = \frac{\sin^2\chi}{r^2}(2r^2 + \sin^2\chi). \qquad (11.54)$$

The spherically symmetric Skyrme equations are obtained by varying L_S with respect to χ. In order to perform variations with respect to the 2–metric $\tilde{g}$, we must take the effective *gravitational*

Lagrangian into account as well. The latter is obtained from the Einstein–Hilbert action after subtracting the appropriate boundary term. A coordinate independent derivation shows that the effective gravitational Lagrangian assumes the form (Brodbeck *et al.* 1996)

$$L_G = -\frac{1}{G}\frac{(dm|dr)}{n}\sqrt{\tilde{g}}.$$ (11.55)

Note that r has an intrinsic meaning defined by the warped structure of a spherically symmetric spacetime (The metric is $^{(4)}g = \pi^\star(\tilde{g}) + (r \circ \pi)^2 \sigma^\star(d\Omega^2)$, where $\pi : M \to \tilde{M} = M/SO(3)$, $\sigma : M \to S^2$ and r is a function on $\tilde{M}$, $r : \tilde{M} \to \mathbb{R}$.) In terms of r, the quantities n and m are defined by

$$n = (dr|dr) = 1 - \frac{2m}{r}.$$ (11.56)

Again using the dimensionless coupling constant $\kappa = 4\pi(f/m_{Pl})^2$, the total effective Lagrangian becomes

$$L = \sqrt{\tilde{g}}\left[-\frac{(dm|dr)}{n} + \frac{\kappa}{2}\{(d\chi|d\chi)u(\chi) + v(\chi)\}\right].$$ (11.57)

Parametrizing the 2–dimensional metric $\tilde{g}$ in terms of the three functions n, s and β,

$$n = (dr|dr), \quad s = \sqrt{\tilde{g}}, \quad \beta = -n^{-1}(dr|dt),$$ (11.58)

the entire set of field equations is now derived by considering variations of L with respect to n, s, β and χ. Since it is always possible to diagonalize $\tilde{g}$, we can set $\beta = 0$ *after* having performed the variations. (Note that β must be introduced in order to obtain the (t,r)–component of Einstein's equations from the effective Lagrangian L. If one is only interested in *static* configurations, one may start by setting $\beta = 0$.) Using $(dm|dr) = n(m' - \beta\dot{m})$, we eventually obtain all field equations by applying the Euler–Lagrange operator D_f^0,

$$D_f^0 = \left[\frac{\partial}{\partial f} - (\frac{\partial}{\partial \dot{f}})^\cdot - (\frac{\partial}{\partial f'})'\right]_{\beta=0},$$

to L. Considering $f = s$, m, β and χ, yields the following set of equations:

$$m' = \frac{\kappa}{2}\left[(n\chi'^2 - \frac{1}{ns^2}\dot{\chi}^2)u + v\right], \qquad (11.59)$$

$$S' = \frac{\kappa}{2}\left[(-\chi'^2 + \frac{1}{n^2 s^2}\dot{\chi}^2)u + v\right], \qquad (11.60)$$

$$\dot{m} = \kappa\left[n\,u\,\dot{\chi}\chi'\right], \qquad (11.61)$$

$$(ns\,u\chi')' - (\frac{u}{ns}\dot{\chi})^{\cdot} = \left(ns\,\chi'^2 - \frac{1}{ns}\dot{\chi}^2\right)\frac{u_\chi}{2} + s\frac{v_\chi}{2}. \qquad (11.62)$$

It is clear that these equations are also obtained from evaluating the general Einstein–Skyrme equations (11.40), (11.41) in the spherically symmetric metric $^{(4)}g = -nsdt^2 + n^{-1}dr^2 + r^2 d\Omega^2$. The above method provides, however, a considerably more efficient derivation, since it does not involve the computation of the stress-energy tensor. For a detailed discussion of the effective action formalism for self–gravitating spherically symmetric fields, we refer to Brodbeck *et al.* (1996).

It is known that the above equations admit static black hole solutions which do not coincide with the Schwarzschild metric (see Luckock and Moss 1986, Droz *et al.* 1991, Heusler *et al.* 1991, 1992, 1993). Since these solutions have a regular event horizon and vanishing asymptotic scalar charge, they provide counter-examples to the "no–hair" conjecture for scalar fields. In fact, the scalar uniqueness theorem, which we shall establish in the next chapter, is restricted to mappings with *harmonic* action. Before we discuss this issue, we will make some comments on another class of scalar fields which also admits nontrivial black hole solutions.

11.5 Conformal scalar fields

The uniqueness theorem mentioned earlier will be established for *minimally coupled harmonic* mappings, that is, the total action will be assumed to have the form ($\kappa = 8\pi G$)

$$S = \int (-\frac{1}{2\kappa}R + e_2[\phi])\eta. \qquad (11.63)$$

We have already pointed out that there do exist static black hole solutions with scalar hair if more general densities, such as the Skyrme Lagrangian, $e_S = e_2 + e_4$, are considered. Another way of generalizing the above action is to consider the additional *non-minimal* coupling term $R \cdot \psi^2$. (Throughout this section ψ denotes a conformally coupled scalar field and the target manifold is assumed to be trivial, $N = \mathbb{R}$.) The conformal action is (Callen *et al.* 1970, Parker 1973)

$$S_{\mathrm{C}} = \int \left[\left(-\frac{1}{2\kappa} + \frac{1}{12}\psi^2 \right) R + \frac{1}{2}(d\psi|d\psi) \right] \eta. \qquad (11.64)$$

Variation with respect to ψ yields the equation

$$\left[\Delta - \frac{1}{6}R \right] \psi = 0. \qquad (11.65)$$

Under a conformal transformation of the spacetime metric, $^{(4)}g \rightarrow g_\Omega = {}^{(4)}g\,\Omega^2$, the Ricci scalar and the Laplacian transform according to

$$R_\Omega = \Omega^{-2}R - 6\Omega^{-3}\Delta\Omega, \qquad (11.66)$$

$$\Delta_\Omega\psi = \Omega^{-2}\left[\Delta\psi + 2\Omega^{-1}(d\psi|d\Omega) \right]. \qquad (11.67)$$

Although neither of these quantities is conformal invariant, the transformation laws imply that $(\Delta_\Omega - \frac{1}{6}R_\Omega)\frac{\psi}{\Omega} = \Omega^{-3}(\Delta - \frac{R}{6})\psi$. Hence, the field equation (11.65) is invariant with respect to transformations

$$^{(4)}g \rightarrow g_\Omega = {}^{(4)}g\,\Omega^2, \qquad \psi \rightarrow \psi_\Omega = \Omega^{-1}\psi. \qquad (11.68)$$

(Note that the numerical value $\frac{1}{12}$ of the conformal coupling factor appearing in the action is crucial. In $n + 1$ dimensions the value is $\frac{n-1}{8n}$.)

We are considering conformally coupled fields mainly because the theory exhibits a static black hole solution with a nontrivial scalar field. This solution, discovered by Bekenstein (1974b, 1975), was the first "counter–example" to Wheeler's famous "no–hair" conjecture. It is, however, important to note that the Bekenstein black hole has the following two flaws: First, the scalar field becomes unbounded on the horizon. Secondly, the horizon is degenerate, that is, the surface gravity vanishes. As was first pointed out by DeWitt (see Bekenstein 1975), the former fact is not actually a severe problem as far as physical implications are concerned (see below). Regarding the horizon, the situation is similar to the electrovac case: Both the Papapetrou–Majumdar (see section 9.5) and the Bekenstein solutions are not covered by the respective uniqueness theorems since they have degenerate horizons. However, in both cases there exist arguments which suggest that these are the only solutions with vanishing surface gravity. (See Xanthopoulos and Zannias 1991 for the uniqueness of the Bekenstein black hole in the spherically symmetric case.)

In the following we adopt the reasoning and notation given by Xanthopoulos and Zannias (1991) and Gal'tsov and Xanthopoulos (1992). In order to complete the set of field equations, one also has to vary the conformal action (11.64) with respect to the metric. This yields Einstein's equations, $G_{\mu\nu} = \kappa T_{\mu\nu}$, where

$$
\begin{aligned}
T_{\mu\nu} &= \psi_\mu \psi_\nu - {}^{(4)}g_{\mu\nu} \frac{1}{2}(d\psi|d\psi) \\
&+ \frac{1}{6}\left[{}^{(4)}g_{\mu\nu}\Delta\psi^2 - \nabla_\mu\nabla_\nu\psi^2 \right] + \frac{1}{6}\psi^2 G_{\mu\nu} . \quad (11.69)
\end{aligned}
$$

As a consequence of the field equations (11.65) and (11.69), the stress–energy tensor is traceless,

$$
\begin{aligned}
\operatorname{tr} T &= -(d\psi|d\psi) + \frac{1}{2}\Delta\psi^2 + \frac{1}{6}\psi^2 \operatorname{tr} G \\
&= \psi\Delta\psi - \frac{1}{6}\psi^2 R = \psi\left(\Delta - \frac{1}{6}R\right)\psi = 0 . \quad (11.70)
\end{aligned}
$$

Hence, provided that Einstein's equations are satisfied, the conformal equation (11.65) reduces to the ordinary equation for a free scalar field. The complete set of field equations for ψ and ${}^{(4)}g$ then becomes

$$
\Delta\psi = 0 , \qquad R_{\mu\nu} = \kappa T^{\mathrm{C}}_{\mu\nu} , \qquad (11.71)
$$

where the conformal stress–energy tensor, $T^{\mathrm{C}}_{\mu\nu}$, is obtained from eq. (11.70), using $G_{\mu\nu} = \kappa T_{\mu\nu}$ and $\Delta\psi = 0$ on the r.h.s.:

$$
T^{\mathrm{C}}_{\mu\nu} = \frac{1}{6 - \kappa\psi^2}\left[4\psi_\mu\psi_\nu - g_{\mu\nu}(d\psi|d\psi) - 2\psi\nabla_\mu\nabla_\nu\psi \right] . \quad (11.72)
$$

Thus, in the presence of Einstein's equations, the distinction between the minimally and the conformally coupled scalar fields lies only in their different energy momentum tensors. (Recall that $T^{\mathrm{h}} = d\phi \otimes d\phi - {}^{(4)}g\frac{1}{2}(d\phi|d\phi)$ for the minimally coupled, *harmonic* scalar field.) By virtue of the matter equation, $\Delta\psi = 0$, the energy momentum tensor is covariantly conserved, $\nabla^\mu T^{\mathrm{C}}_{\mu\nu} = 0$, which is also required by the Bianchi identity for $G_{\mu\nu}$ and Einstein's equations. (This can also be verified by using eq. (1.4).)

A very useful tool for generating solutions of the field equations (11.71) was given by Bekenstein (1974b):

Theorem 11.11 *Let* $(\phi, {}^{(4)}g)$ *be a solution of the minimally coupled harmonic scalar field equations,* $\Delta\phi = 0$, $G_{\mu\nu} = \kappa T^h_{\mu\nu}$. *Then* $(\psi, {}^{(4)}\hat{g})$ *is a solution of the conformally coupled equations (11.71), if*

$$\psi = \sqrt{\frac{6}{\kappa}} \tanh\left(\sqrt{\frac{\kappa}{6}}\phi\right), \qquad (11.73)$$

$$^{(4)}\hat{g} = \left(1 - \frac{\kappa}{6}\psi^2\right)^{-1} {}^{(4)}g = \cosh^2\left(\sqrt{\frac{\kappa}{6}}\phi\right) {}^{(4)}g. \qquad (11.74)$$

The proof of this theorem is obtained from the transformation law for the Einstein tensor under conformal transformations (see, e.g., Wald 1984). Bekenstein's original theorem is even more general, in the sense that electromagnetic fields are also included.

We now derive the Bekenstein black hole solution (Bekenstein 1974b). We do so by first constructing a family of solutions to the minimally coupled free scalar field equations, using a generating technique due to Janis *et al.* (1969). Despite the fact that the solutions obtained in this way are physically not acceptable, one can use them to obtain a family of solutions to the *conformal* equations by applying the above theorem. In order to find a solution of the *minimal* scalar field equations, one writes the static metric (2.38) in the form

$$^{(4)}g = -S^2\, dt^2 + g = -S^2\, dt^2 + S^{-2}\, h. \qquad (11.75)$$

Under the 3–dimensional conformal transformation $h = S^2 g$, the Laplacian and the Ricci tensor on Σ become, respectively,

$$\Delta^{(g)} S = S^3 \Delta^{(h)} \ln S, \qquad (11.76)$$

$$R^{(g)}_{ij} - S^{-1}\nabla^{(g)}_i \nabla^{(g)}_j S$$

$$= R^{(h)}_{ij} + h_{ij}\Delta^{(h)} \ln S - 2\nabla^{(h)}_i \ln S\, \nabla^{(h)}_j \ln S. \qquad (11.77)$$

Using Einstein's equations for a static, minimally coupled scalar field, $R_{tt} = 0$, $R_{ij} = \kappa\phi_i\phi_j$, the general expressions (2.45), (2.46) now yield

$$\Delta^{(h)} \ln S = 0, \qquad (11.78)$$

$$R^{(h)}_{ij} = \kappa\,(\,\phi_i\phi_j + 2\nabla_i \ln S \nabla_j \ln S\,), \qquad (11.79)$$

$$\Delta^{(h)}\phi = 0. \qquad (11.80)$$

The last equation arises from the fact that the 4–dimensional Laplacian of a static field coincides with the 3–dimensional Laplacian with respect to the metric h.

The observation of Janis *et al.* (1969) comprises the fact that the pair (S, ϕ),

$$S = (S_0)^\alpha, \qquad \phi = \sqrt{\frac{2}{\kappa}} \sqrt{1 - \alpha^2} \, \ln S_0, \qquad (11.81)$$

solves the above equations if $(S_0, \phi_0 = 0)$ is a vacuum solution and α is a real constant, $\alpha \in [0, 1]$.

The Bekenstein black hole is now obtained by applying both generating techniques to the Schwarzschild vacuum solution. In terms of S_0 and h, the latter reads

$$S_0 = \sqrt{1 - \frac{2m}{r}}, \qquad h = dr^2 + (S_0)^2 \, r^2 \, d\Omega^2. \qquad (11.82)$$

By virtue of the above formulae, this yields the following 1–parameter family of minimally coupled scalar field solutions:

$$\phi = \sqrt{\frac{2}{\kappa}} \sqrt{1 - \alpha^2} \, \ln S_0, \qquad (11.83)$$

$$^{(4)}g = -(S_0)^{2\alpha} dt^2 + (S_0)^{-2\alpha} dr^2 + (S_0)^{2(1-\alpha)} r^2 \, d\Omega^2. \qquad (11.84)$$

Before we use this solution to generate the Bekenstein black hole, we note the following: For $\alpha \in [0, 1]$, the above solutions are asymptotically flat and have non–negative total mass, $M = \alpha m$. (This is immediately seen from the Komar formula, using $*dk = -[(S_0)^{2\alpha}]'(S_0)^{2(1-\alpha)} r^2 d\Omega^2$.) Using $R_{00} = 0$ and $R_{ij} = \kappa \phi_i \phi_j$, one finds that (for all admissible values of the parameter α) the curvature invariant $R_{\mu\nu} R^{\mu\nu}$ becomes unbounded for $r = 2m$, unless $\alpha = 1$. In fact, for all $\alpha \in [0, 1[$, $r = 2m$ is a naked singularity. The exception $\alpha = 1$ is, of course, the Schwarzschild vacuum solution, $S = S_0$, $\phi = 0$.

In order to obtain solutions to the *conformal* scalar field equations, we finally apply the transformation (11.73), (11.74). The above family then yields the following solutions:

$$^{(4)}\hat{g} = \frac{1}{4} \left[(S_0)^\gamma + (S_0)^{-\gamma} \right]^2 {}^{(4)}g, \qquad (11.85)$$

$$\psi = \sqrt{\frac{6}{\kappa}} \frac{1 - (S_0)^{2\gamma}}{1 + (S_0)^{2\gamma}}, \qquad \text{with } \gamma = \sqrt{\frac{1}{3}(1 - \alpha^2)}, \qquad (11.86)$$

where $^{(4)}g$ is the metric (11.84). Computing the invariant $R_{\mu\nu}R^{\mu\nu}$ for this solution shows that the surface $r = 2m$ corresponds again to a naked curvature singularity (see Bekenstein 1974b for a discussion). However, in addition to the exception $\alpha = 1$, corresponding to the Schwarzschild solution, there now exists a second value for which the solution remains regular: For $\alpha = \frac{1}{2}$ ($\gamma = \frac{1}{2}$) the surface $r = 2m$ is nonsingular and has the finite area πm^2. Introducing the new radial coordinate $\rho(r) = m[1 - S_0(r)]^{-1}$, the metric (11.85) assumes the form ($\alpha = \gamma = \frac{1}{2}$)

$$^{(4)}\hat{g} = -\left(1 - \frac{m}{2\rho}\right)^2 dt^2 + \left(1 - \frac{m}{2\rho}\right)^{-2} d\rho^2 + \rho^2 d\Omega^2, \quad (11.87)$$

which is completely regular for $r = 2m$ (i.e. $\rho = m$). However, there exists a horizon for $\rho = \frac{m}{2}$. Since the norm of the timelike Killing field is $N = -(1 - \frac{m}{2\rho})^2$, we find that dN vanishes on the horizon, implying that $\kappa = 0$ (see eq. (6.14)). In fact, the above metric describes exactly the Reissner–Nordström geometry with an extreme horizon, i.e., with $|Q| = M$.

In terms of the new coordinate ρ, the conformal scalar field becomes $\psi = \sqrt{\frac{6}{\kappa}}(\frac{2\rho}{m} - 1)^{-1}$, which is unbounded on the horizon. Nevertheless, the above metric describes a smooth, asymptotically flat geometry with finite mass $\frac{m}{2}$. Any test objects interacting only gravitationally do not feel the unbounded horizon magnitude of the conformal field. In addition, for test particles which *do* interact with the conformal field, the horizon occurs at an infinite proper time (Bekenstein 1975).

In the spherically symmetric case, the uniqueness of the Bekenstein solution was settled by Xanthopoulos and Zannias (1991). In view of the uniqueness theorem for harmonic maps we also draw the following conclusion from the above example: Any finiteness assumptions for scalar fields which cannot be derived from geometrical requirements - such as asymptotic flatness, boundedness of curvature invariants or regularity properties of the event horizon - must be handled very carefully. We shall encounter this problem in section 12.4, where a discussion of boundary terms will be necessary to exclude nontrivial harmonic fields.

12

Self–gravitating harmonic mappings

In this chapter we establish the uniqueness of the Kerr metric amongst the stationary black hole solutions of self–gravitating *harmonic* mappings (scalar fields) with arbitrary Riemannian target manifolds. As in the vacuum and the electrovac cases, the uniqueness proof consists of three main parts: First, taking advantage of the strong rigidity theorem (see section 6.2), one establishes staticity for the nonrotating case, and circularity for the rotating case. One then separately proves the uniqueness of the Schwarzschild metric amongst all static configurations, and the uniqueness of the Kerr metric amongst all circular black hole solutions.

The three problems mentioned above are treated in the first section and the last two sections, respectively: The staticity and circularity theorems are derived from the symmetry properties of the scalar fields and the general theorems given in sections 8.1 and 8.2. The static uniqueness theorem is then proven along the same lines as in the vacuum case, that is, by means of conformal techniques and the positive energy theorem. The uniqueness theorem for rotating configurations turns out to be a consequence of the corresponding vacuum theorem (see chapter 10) and an additional integral identity for stationary and axisymmetric harmonic mappings.

Besides dealing with general harmonic mappings, we shall also pay some attention to ordinary scalar (Higgs) fields. By this, we mean harmonic mappings into linear target spaces with an additional potential term in the Lagrangian. By 1972, Bekenstein had already established the static no–hair theorem for ordinary massive scalar fields, by means of a divergence identity. Using a Bochner (1940) identity, it is not hard to extend the argument to models with a *convex* potential (see the fourth section). How-

ever, without the convexity requirement, the problem is still open
- unless additional symmetries are imposed: In fact, in the spherically symmetric case, there exist at least three different arguments
which exclude black hole solutions with arbitrary non–negative
potentials (see Heusler 1992, 1995a, Heusler and Straumann 1992
and Sudarsky 1995).

We address the above issues in the remaining parts of this chapter: In the second section we take advantage of the Komar formula
to construct a contradiction to the positive energy theorem for
self–gravitating *soliton* solutions of field theories with $R(k, k) \leq 0$
(where k denotes the timelike Killing field). This immediately
yields an efficient proof of the uniqueness theorem for harmonic
soliton mappings with arbitrary non–negative potentials and arbitrary Riemannian target manifolds. The third section deals with
a modification of this reasoning, which provides the corresponding
uniqueness theorem for spherically symmetric *black hole* solutions
(Heusler 1995a).

The "traditional" approach to the static uniqueness theorem,
based on integral identities (Bekenstein 1972), is presented in the
fourth section. It yields positive results for harmonic mappings
with *convex* potentials and target manifolds with *nonpositive* sectional curvature.

As already mentioned, the difficulties concerning the curvature
of the target manifold are eliminated in the uniqueness proof for
self–gravitating harmonic mappings which will be presented in the
last two sections (Heusler 1993). If the potential is not assumed to
be convex, the problem remains open, unless spherical symmetry
is imposed. (We note that solutions with nonconvex potentials
have been found numerically (Bechmann and Lechtenfeld 1995).)
However, their existence is probably less due to the fact that the
potential has local maxima, than to the circumstance that the
model violates the dominant energy condition in those parts of
the domain where the potential is negative.

12.1 Staticity and circularity

Throughout this chapter, the terms *general* and *ordinary* scalar
field will be used in order to distinguish between mappings into
arbitrary Riemannian manifolds, N, and mappings into $\mathbb{R}^n$.

By virtue of the strong rigidity theorem, the asymptotically timelike Killing field either coincides with the generator of the horizon or the spacetime is stationary *and* axisymmetric. In both cases one must establish the appropriate integrability conditions in order to write the metric in the static or circular form. We do so by introducing the notions of stationary and axisymmetric mappings and applying the general integrability theorems given in sections 8.1 and 8.2.

Definition 12.1 *A mapping (general scalar field)* $\phi : (M, g) \to$ (N, G) *is called stationary, if it is invariant under the 1–parameter group of transformations generated by the asymptotically timelike Killing field* k, $L_k \phi = 0$. *If* ϕ *is invariant under the 1–parameter group generated by the axial Killing field* m, $L_m \phi = 0$, *then the mapping is called axisymmetric.*

In order to apply the staticity and circularity theorems, we have to compute the energy–momentum 1–forms $T(k)$ and $T(m)$ associated with ϕ. If the matter Lagrangian, $\mathcal{L} = \mathcal{L}(g, G, \phi, d\phi)$, does not depend on derivatives of the spacetime metric, then the stress–energy tensor becomes

$$T = 2 \frac{\partial \mathcal{L}}{\partial g^{-1}} - \mathcal{L} \cdot g = f_{AB} \, d\phi^A \otimes d\phi^B - \mathcal{L} \cdot g, \qquad (12.1)$$

where the $f_{AB} = f_{AB}(g, G, \phi, d\phi))$ depend on the explicit form of the Lagrangian. (In the harmonic case, where $\mathcal{L} = e_2$, we have $f_{AB} = G_{AB}$, whereas for the Skyrme density, $\mathcal{L} = e_2 + e_4$, we obtain $f_{AB} = G_{AB} + (G_{AB}G_{CD} - G_{AD}G_{BC})(d\phi^C | d\phi^D)$.) If ϕ is invariant with respect to the action of the Killing field K (where $K = k$ or $K = m$), then we have $i_K d\phi = L_K \phi = 0$. The stress–energy 1–form with respect to K becomes proportional to K itself,

$$T(K) = -\mathcal{L} \cdot K. \qquad (12.2)$$

This implies, of course, that stationary mappings satisfy the condition

$$k \wedge T(k) = 0, \qquad (12.3)$$

whereas stationary and axisymmetric mappings fulfil

$$m \wedge k \wedge T(k) = 0, \qquad k \wedge m \wedge T(m) = 0. \qquad (12.4)$$

Since these are the Ricci conditions (8.1) and (8.2), respectively, we can apply the theorems 8.1 and 8.2 to draw the following conclusions:

Corollary 12.2 *Consider an asymptotically flat spacetime with nondegenerate, nonrotating (not necessarily connected) horizon and strictly stationary domain of outer communications* $\ll S \gg$. *Let* ϕ *be a self-gravitating stationary mapping from* $\ll S \gg$ *into a Riemannian manifold. Then* $\ll S \gg$ *is static.*

Corollary 12.3 *Consider an asymptotically flat spacetime with nondegenerate, rotating horizon and stationary and axisymmetric* $\ll S \gg$. *Let* ϕ *be a self-gravitating stationary and axisymmetric mapping from* $\ll S \gg$ *into a Riemannian manifold. Then* $\ll S \gg$ *is circular.*

In view of the uniqueness proof for static black holes, it is crucial to note that for an invariant mapping, $L_K \phi = 0$, with a *harmonic* density, the Ricci 1–form $R(K)$ vanishes identically:

$$\mathcal{L} = e_2[\phi] \implies T(K) - \frac{1}{2}K \operatorname{tr} \boldsymbol{T} = 0 . \qquad (12.5)$$

This observation - together with the positive energy theorem on the one hand, and the Komar mass expression on the other - enables one to prove a nonexistence theorem for asymptotically flat, self-gravitating harmonic *soliton* solutions. In fact, there are good reasons to conjecture that the violation of the *strong* energy condition, $R(k, k) \leq 0$, also prohibits the existence of static black holes other than the Schwarzschild solution in any field theory satisfying the *dominant* energy condition. Under the additional assumption of spherical symmetry, this can indeed be proven, as we shall see in the next section.

We finally point out that the vanishing of the Ricci 1–form is a feature particular to the *harmonic* density. As an example, we obtain for the Skyrme Lagrangian, $e_S = e_2 + e_4$,

$$\mathcal{L} = e_S[\phi] \implies T(K) - \frac{1}{2}K \operatorname{tr} \boldsymbol{T} = -e_4[\phi] K . \qquad (12.6)$$

This also shows that, in contrast to the harmonic case, the strong energy condition *does* hold for all Skyrme mappings with strictly stationary domain, $R(k, k) = -\kappa\, e_4[\phi]\, (k|k) \geq 0$.

12.2 Nonexistence of soliton solutions

As we have argued above, all stationary *harmonic* mappings have vanishing Ricci 1–form with respect to the Killing field k, also implying $R(k, k) = 0$. If, in addition to the harmonic density, $e_2[\phi]$, the Lagrangian also contains a *non–negative* potential term, $P[\phi]$, the situation becomes even more constrained, in the sense that the strong energy condition is strictly violated whenever k is not null and P does not vanish:

$$\mathcal{L} = e_2[\phi] + P[\phi] \implies$$

$$T(k, k) - \frac{1}{2}(k|k)\,\mathrm{tr}\,\boldsymbol{T} = P[\phi]\,(k|k) \leq 0. \tag{12.7}$$

The first aim of this section is to derive an expression for the Komar mass involving the twist of k and $R(k, k)$. We then apply this formula to strictly stationary spacetimes without horizons. The positive energy theorem then enables us to establish a general uniqueness result for *soliton* solutions of matter models which violate the strong energy condition, such as harmonic mappings with non–negative potentials. (Self–gravitating soliton solutions can exist in theories which fulfil the strong energy condition, such as Einstein–Yang–Mills models; see Bartnik and McKinnon 1988.)

Consider the Komar mass (2.35) of an asymptotically flat stationary spacetime with nonrotating horizon,

$$M = \frac{1}{4\pi}\kappa\mathcal{A} - \frac{1}{4\pi}\int_\Sigma *R(k). \tag{12.8}$$

Taking advantage of the general identity (2.16) for the derivative of the twist $\omega \equiv \omega_k$, we have (with $V = -N = -(k|k)$)

$$k \wedge d\omega = V * R(k) + R(k, k) * k, \tag{12.9}$$

by means of which we can express the integrand in eq. (12.8) in terms of $R(k, k)$:

$$M = \frac{1}{4\pi}\kappa\mathcal{A} + \frac{1}{4\pi}\int_\Sigma \frac{R(k, k)}{V} * k - \frac{1}{4\pi}\int_\Sigma \frac{k \wedge d\omega}{V}. \tag{12.10}$$

In order to proceed, we integrate the general identity 8.4 over Σ. As we argued in the proof of the staticity theorem 8.2, the boundary term does not contribute. We therefore have

$$\int_\Sigma \frac{k \wedge d\omega}{V} = 2\int_\Sigma \frac{(\omega|\omega)}{V^2} * k. \tag{12.11}$$

Hence, we obtain the following useful expression for the Komar mass of a stationary spacetime with nonrotating Killing horizon:

$$M = \frac{1}{4\pi} \kappa \mathcal{A} + \frac{1}{4\pi} \int_\Sigma \frac{R(k,k)}{V} i_k \eta - \frac{1}{2\pi} \int_\Sigma \frac{(\omega|\omega)}{V^2} i_k \eta. \quad (12.12)$$

If the domain of outer communications is strictly stationary, then k is nowhere spacelike and ω is nowhere timelike. In this case, the violation of the strong energy condition implies an *upper* bound for the total mass in terms of the horizon quantities κ and $\mathcal{A}$:

$$M \leq \frac{1}{4\pi} \kappa \mathcal{A}, \quad \text{if} \quad T(k,k) + \frac{1}{2} V \, \text{tr} \, \mathbf{T} \leq 0. \quad (12.13)$$

Restricting ourselves to spacetimes *without* horizons, the above inequality reduces to $M \leq 0$. On the other hand, the positive mass theorem yields $M \geq 0$ as a consequence of the *dominant* energy condition. We thus obtain the following uniqueness theorem for *soliton* solutions of a class of self–gravitating field theories (Heusler 1995a):

Theorem 12.4 *Let $(M, \boldsymbol{g})$ be a strictly stationary, asymptotically flat spacetime coupled to a matter model satisfying the dominant energy condition and violating the strong energy condition for the Killing field k at every point, $R(k, k) \leq 0$. Then $(M, \boldsymbol{g})$ is flat.*

In order to apply this result to scalar fields, we consider Higgs fields on the one hand and harmonic mappings on the other. (We use the term Higgs field for a mapping into a *vector* space with Lagrangian $e_2[\phi] + P[\phi]$.) If the potential, $P[\phi]$, is non–negative, then the conclusion $M = 0$ immediately yields $P[\phi] = 0$, implying that ϕ is a constant map. For a harmonic mapping into an arbitrary Riemannian manifold, the facts that the base manifold must be flat and the target manifold is Riemannian also show that ϕ is constant. We thus obtain the following two corollaries to the above theorem:

Corollary 12.5 *Let $(M, \boldsymbol{g})$ be a strictly stationary, asymptotically flat spacetime coupled to a stationary Higgs field ϕ with arbitrary non–negative potential. Then $(M, \boldsymbol{g})$ is flat and ϕ is a zero of the potential.*

Corollary 12.6 *Let (M, g) be a strictly stationary, asymptotically flat spacetime coupled to a harmonic mapping ϕ with arbitrary Riemannian target manifold (N, G). Then (M, g) is flat and ϕ is a constant map.*

It is, of course, tempting to try to extend the above reasoning to black hole spacetimes. This is indeed possible if the inequality $M \geq \frac{1}{(4\pi)}\kappa\mathcal{A}$ can be obtained as a consequence of the *dominant* energy condition. This estimate has, however, not been established yet. In fact, the Penrose (1973) conjecture, $M^2 \geq \mathcal{A}/(16\pi)$, which is closely related to the above inequality, has been proven only under additional geometrical assumptions (Ludvigson and Vickers 1983) or in the spherically symmetric case (see also Geroch 1973, Jang and Wald 1977, Jang 1978 and Heusler 1995a). In the general case, the strongest result consists in the generalization of Witten's proof of the positive mass theorem (Witten 1981; see also Parker and Taubes 1982) by Gibbons *et al.* (1983), establishing $M \geq 0$ also for *black hole* spacetimes.

As we shall see in the next section it is, however, not hard to derive the desired bound in the spherically symmetric case. This yields the conclusion that all static, spherically symmetric black hole solutions with matter satisfying the dominant energy condition, but violating the strong energy condition (for k), coincide with the Schwarzschild metric.

12.3 Uniqueness of spherically symmetric black holes

In the strictly stationary case all black hole solutions of matter models with $R(k, k) \leq 0$ are subject to the inequality (12.13). The purpose of this section is to establish the converse inequality on the basis of the *dominant* energy condition. However, for the time being, this seems to be possible only under the additional assumption of spherical symmetry.

A static and spherically symmetric metric is parametrized in terms of two functions of the radial coordinate r. Using the quantities $m(r)$ and $s(r)$ (see section 11.4), we have

$$g = -n\,s^2\,dt^2 + n^{-1}\,dr^2 + r^2\,d\Omega^2\,, \tag{12.14}$$

$$n(r) = 1 - \frac{2m(r)}{r}\,. \tag{12.15}$$

(In terms of the familiar Schwarzschild parametrization one has $s = e^{(a+b)}$ and $n = e^{-2b}$.) Tensor components refer to the orthonormal frame of 1–forms $\theta^0 = \sqrt{ns}\,dt$, $\theta^1 = n^{-1/2}dr$, $\theta^2 = r\,d\vartheta$, $\theta^3 = r\sin\vartheta\,d\varphi$, where s and n are positive, except on the horizon, $n(r_H) = 0$. In terms of these quantities, the Killing 1–form becomes $k = -ns^2 dt = -\sqrt{ns}\,\theta^0$. Since $*(\theta^0 \wedge \theta^1) = -\theta^2 \wedge \theta^3$, this yields $*dk = s^{-1}(ns^2)' * (\theta^0 \wedge \theta^1) = -r^2 s^{-1}(ns^2)'d\Omega$. Defining the "local mass" $M(r)$ by the Komar integral over the 2–sphere S_r^2 with coordinate radius r, we find

$$M(r) = -\frac{1}{8\pi}\int_{S_r^2} *dk = \frac{r^2}{2S}(ns^2)' = ms + ns'r^2 - m'sr \,. \quad (12.16)$$

As we have argued before, $M(r)$ is a decreasing function if $R(k,k)$ is negative, that is, if the strong energy condition is violated. In contrast to this, we shall now establish that $\lim_\infty M(r) \geq M_H \equiv M(r_H)$ if the *dominant* energy condition holds.

With respect to the orthonormal frame $\{\theta^\mu\}$ we have

$$G_{00} = \frac{2}{r^2}\,m' \,, \qquad G_{00} + G_{11} = \frac{2n}{r}\frac{s'}{s} \,. \quad (12.17)$$

Requiring that $-T(X)$ is future directed timelike or null for all future directed timelike vectors X, Einstein's equations imply that the quantities G_{00} and $G_{00} + G_{11}$ are non–negative. Hence, provided that matter is subject to the dominant energy condition, both functions, $m(r)$ and $s(r)$, are increasing,

$$m'(r) \geq 0 \,, \qquad s'(r) \geq 0 \quad \text{for } r \geq r_H \,. \quad (12.18)$$

In order to derive estimates for $M(r)$ at infinity and at the horizon, it remains to note that the last term in eq. (12.16) does not contribute for $r \to \infty$, whereas the second term vanishes on the horizon. We thus obtain an upper bound for M_H and a lower bound for $M \equiv \lim_\infty M(r)$. More precisely, asymptotic flatness implies the existence and finiteness of $m_\infty \equiv \lim_\infty m(r)$, $s_\infty \equiv \lim_\infty s(r)$ and $\lim_\infty(r^2 s')$. Since, in addition, $\lim_\infty(rm')$ vanishes and the second term in eq. (12.16) is non–negative, we have $M \geq m_\infty s_\infty$. The facts that $n(r_H) = 0$ and $m'sr$ is non–negative yields the estimate $M_H \leq m_H S_H$ for the Komar integral at the horizon. Hence, again using $m' \geq 0$ and $s' \geq 0$, we eventually find

$$M_H \leq m_H s_H \leq m_\infty s_\infty \leq M. \qquad (12.19)$$

This shows that the entire Komar integral is non–negative if the matter fields are subject to the dominant energy condition,

$$-\frac{1}{8\pi} \int_{\partial\Sigma} *dk = M - M_H = M - \frac{1}{4\pi}\kappa\mathcal{A} \geq 0, \qquad (12.20)$$

where $\partial\Sigma = S_\infty^2 - \mathcal{H}$.

We are now able to conclude that both $m(r)$ and $s(r)$ assume constant values if the dominant energy condition holds but the strong energy condition is violated for the timelike Killing field. In this case, the metric (12.14) coincides with the Schwarzschild metric and both T_{00} and $T_{00} + T_{11}$ must vanish. We thus have the following uniqueness result for harmonic mappings and Higgs fields (Heusler 1995a):

Corollary 12.7 *Let (M, g) be a spherically symmetric, asymptotically flat static black hole spacetime, and let ϕ be a self–gravitating stationary harmonic mapping with Riemannian target manifold, either with or without an arbitrary non–negative potential. Then g is the Schwarzschild metric and ϕ is a trivial map with zero energy density.*

The uniqueness of the Schwarzschild metric amongst the spherically symmetric black hole solutions with scalar fields with arbitrary non–negative potentials was also proven by using a generalized version of scaling arguments (Heusler 1992, Heusler and Straumann 1992). A different proof, based on the construction of a monotonic function, was presented by Sudarsky (1995) for Higgs fields. The argument presented above does not use any details of the matter model. It simply plays the strong and the dominant energy conditions against each other.

It is important to be aware of the fact that $R(k, k) \leq 0$ is special to *minimally* coupled *harmonic* mappings (with arbitrary Riemannian target space and arbitrary non–negative potentials). Hence, neither the Skyrme model nor the conformally coupled scalar field are covered by the above reasoning. In fact, both of these theories *do* admit spherically symmetric black hole solutions "with hair" (see the previous chapter).

To conclude, we observe the following consequences of the above discussion: The dominant energy condition together with the expression (12.16) provides a simple estimate for the surface gravity of a spherically symmetric black hole: Evaluating the formula for $M(r)$ at the horizon and using $M_H = \frac{1}{4\pi}\kappa\mathcal{A} = \kappa r_H^2$ immediately yields

$$\kappa = \frac{s_H}{2r_H}(1 - 2m_H'). \tag{12.21}$$

Since $m'(r) \geq 0$ and $s_H \leq s_\infty = 1$, we recover the fact (Visser 1992) that the Hawking temperature $T_H = \kappa\hbar/(2\pi k)$ of a nondegenerate, spherically symmetric black hole with matter satisfying the dominant energy condition is bounded from above by the Hawking temperature $T_H^{(vac)}$ of the Schwarzschild black hole with the same area,

$$T_H = \frac{\hbar}{2\pi k}\frac{s_H}{2r_H}(1 - 2m_H') \leq \frac{\hbar}{2\pi k}\frac{1}{2r_H} = T_H^{(vac)}. \tag{12.22}$$

We also point out that the inequality (12.20) (positivity of the Komar boundary integral) is weaker than the well-known Penrose (1973) conjecture

$$M^2 \geq \frac{\mathcal{A}}{16\pi}. \tag{12.23}$$

However, in the static and spherically symmetric case under consideration, the latter is also established from eq. (12.19) since (using $m' \geq 0$ and the gauge $s_\infty = 1$) one has $M \geq m_H = r_H/2 = \sqrt{\mathcal{A}/(16\pi)}$. We finally emphasize that eq. (12.21), together with $s_H \leq s_\infty = 1$ and $m' \geq 0$, yields the inequality $\kappa \leq \sqrt{\pi/\mathcal{A}}$. Using this and the Penrose inequality, we find

$$M \geq \sqrt{\frac{\mathcal{A}}{16\pi}} \geq \frac{\kappa}{4\pi}\mathcal{A}. \tag{12.24}$$

The inequality between the surface gravity and the area of the horizon is also obtained in the nonstatic case, applying the generalized version of the zeroth law (Hayward 1994a). In addition, Hayward (1994b, 1994d) also established the Penrose conjecture (12.23) for the nonstatic, spherically symmetric case, using the Misner–Sharp (1964) local energy.

12.4 Divergence identities

In 1972 Bekenstein showed that there are no black holes with nontrivial massive scalar fields. Basically, his strategy consisted of multiplying the matter equation $(\Delta - m^2)\phi = 0$ by ϕ, integrating by parts and using Stokes' theorem for the resulting identity

$$\frac{1}{\sqrt{g}}(\sqrt{g}\phi^\mu)_{,\mu} = \nabla^\mu\phi\nabla_\mu\phi + m^2\phi^2. \qquad (12.25)$$

By arguing that the boundary integral vanishes on both parts of $\partial\Sigma$, S^2_∞ and $\mathcal{H} = H \cap \Sigma$, Bekenstein was able to conclude that a static massive scalar field ϕ must vanish in the domain of outer communications.

The aim of this section is to discuss this reasoning in some detail and to extend it to the largest possible class of potentials $P[\phi]$ on the one hand, and to the largest possible class of target manifolds on the other. In view of Bekenstein's black hole solution presented in section 11.5, we shall give some emphasis to the discussion of the boundary terms.

Using divergence identities, it turns out that uniqueness results for static black hole solutions can only be derived for mappings with *convex* potentials, into target manifolds with *negative* sectional curvature. Regarding the convexity of the potential, there exists no better result until now. The restriction on the curvature of the target manifold can, however, be removed by using different methods (see the next section). We also recall that both problems are absent in the spherically symmetric case (see the previous section).

In the first part of this section we consider minimally coupled ordinary scalar fields, $\phi : M \to \mathbb{R}$, with non–negative potentials $P[\phi]$ (Higgs fields),

$$S = \int \left(-\frac{1}{2\kappa}R + \frac{1}{2}(d\phi|d\phi) + P[\phi]\right)\eta. \qquad (12.26)$$

Variation with respect to ϕ and g yields the matter and Einstein equations, respectively,

$$\Delta\phi = P_\phi, \qquad (12.27)$$

$$R_{\mu\nu} = \kappa\left(\phi_{,\mu}\,\phi_{,\nu} + g_{\mu\nu}P[\phi]\right), \qquad (12.28)$$

where Δ denotes the ordinary spacetime Laplacian, $\Delta\phi = -d^\dagger d\phi = \frac{1}{\sqrt{g}}(\sqrt{g}g^{\mu\nu}\phi_{,\nu})_{,\mu}$.

Integral identities are obtained from Stokes' theorem which, for arbitrary stationary 1–forms α ($L_k\alpha = 0$), reads

$$\int_{\partial\Sigma} *(k \wedge \alpha) = -\int_\Sigma d^\dagger\alpha \, i_k\eta. \qquad (12.29)$$

As usual, Σ is the spacelike hypersurface extending from the inner boundary $\mathcal{H} = \Sigma \cap H$ to S^2_∞. In the following we shall assume that the domain of outer communications is *strictly* static. Let $f[\phi]$ be an arbitrary expression in ϕ and let

$$\alpha = f[\phi] \, d\phi, \qquad (12.30)$$

where stationarity of ϕ implies $L_k\alpha = 0$. Using $d^\dagger\alpha = -(df|d\phi) + f d^\dagger d\phi = -f_\phi(d\phi|d\phi) - \Delta\phi$ and the matter equation (12.27), we obtain from Stokes' theorem

$$\int_{\partial\Sigma} f[\phi] * (k \wedge d\phi) = \int_\Sigma [f_\phi(d\phi|d\phi) + f \, P_\phi] \, i_k\eta. \qquad (12.31)$$

(For $f = 1$ and $P = \frac{1}{2}m^2\phi^2$ we recover the identity (12.25).) The aim is now to choose $f[\phi]$ such that the boundary integral vanishes and, in addition, the integrand on the r.h.s. has a fixed sign. We start by focusing on the horizon contribution to the boundary term: Using the general formula (7.15) to compute the integral of a 2–form on the horizon, we find

$$\int_{\mathcal{H}} *(k \wedge d\phi) = \int_{\mathcal{H}} i_n i_k (k \wedge d\phi) \, d\mathcal{A} = \int_{\mathcal{H}} (k|k) \, L_n\phi = 0.$$

Here we have used $L_k\phi = 0$ and the fact that k is null on the horizon. Hence, the boundary term from the horizon does not contribute, provided that $f[\phi]$ remains finite on H. Turning to the integral over S^2_∞, asymptotic flatness requires that the scalar charge density, $*(k \wedge d\phi)$, does not become infinite and vanishes only if ϕ decays faster than r^{-1}. Thus, in order to avoid overly restrictive assumptions on the asymptotic behavior of the scalar field, we must consider asymptotically vanishing functions $f[\phi]$. To summarize, we have to find $f[\phi]$ such that

$$|f[\phi]| < \infty \text{ on } H, \qquad f[\phi] = 0 \text{ on } S^2_\infty. \qquad (12.32)$$

We are now in the position to prove the following uniqueness theorem:

Theorem 12.8 *Let ϕ be a stationary, minimally coupled self-gravitating scalar field, $\phi : M \rightarrow \mathbb{R}$, with harmonic density and non-negative convex potential $P[\phi]$ (with $P[\phi] < \infty$ for $|\phi| < \infty$). Then the only black hole solution of the field equations (12.27) and (12.28) with regular and nonrotating event horizon and strictly stationary, asymptotically flat domain of outer communications is the Schwarzschild metric with a constant field ϕ_0 with $P[\phi_0] = 0$.*

Proof First, the staticity theorem for stationary scalar fields implies that the domain of outer communications is static if the horizon is nonrotating. We now show that the choice

$$f[\phi] = P_\phi[\phi] \qquad (12.33)$$

(i.e., $\alpha = dP$; see eq. (12.30)) satisfies the requirements (12.32): Since the curvature invariant

$$R_{\mu\nu}R^{\mu\nu} = \kappa^2 \left\{ \left[(d\phi|d\phi) + P \right]^2 + 3P^2 \right\}$$

must remain finite, we conclude that the potential, and hence ϕ and also P_ϕ, remain finite on the horizon. (Note that this part of the argument does not work for free or conformally coupled scalar fields. In fact, the scalar field of the Bekenstein black hole solution presented in section 11.5 becomes infinite on the horizon, even though the above invariant remains bounded.) Since the Komar formula can be written as an integral of the potential over Σ (see eq. (12.8)), asymptotic flatness requires that $P[\phi]$ vanishes on S_∞^2. The assumption $P[\phi] \geq 0$ then implies that, asymptotically, $\phi \rightarrow \phi_0$ with $P_\phi[\phi_0] = 0$ (where ϕ_0 is not required to be finite). Hence, $f = P_\phi$ fulfils the conditions (12.32) from which, together with eq. (12.31), we conclude that

$$\int_\Sigma [P_{\phi\phi}(d\phi|d\phi) + P_\phi^2] i_k \eta = 0.$$

Since the domain is strictly static, $(k|d\phi) = L_k\phi = 0$ implies that $d\phi$ is nowhere timelike. Together with the convexity of P, this yields the conclusion that both integrands vanish separately. Hence, $\phi = \phi_0$, and the field equations reduce to the vacuum equations. Applying the vacuum uniqueness theorem for the Schwarzschild solution then concludes the proof. $\qquad\square$

As mentioned above, this proof applies to free scalar fields only if ϕ is *assumed* to remain finite on the horizon. (The finiteness of the curvature invariant does not imply the finiteness of ϕ in the absence of the potential.) However, the case without potential is covered by the following theorem, which also applies to harmonic mappings into nonpositively curved target manifolds (see also Gibbons 1991):

Theorem 12.9 *Let ϕ be a stationary, minimally coupled, self-gravitating harmonic mapping from spacetime into a Riemannian manifold (N, G) with nonpositive sectional curvature. Then the only black hole solution with regular and nonrotating event horizon and strictly stationary, asymptotically flat domain of outer communications is the Schwarzschild metric with a constant mapping ϕ_0.*

Proof We use again the static Stokes formula (12.29): Instead of considering $\alpha = d P[\phi]$, we now choose $\alpha = d e_2[\phi]$. Stokes' theorem now yields

$$\int_{\partial \Sigma} * (k \wedge de_2[\phi]) = \int_{\Sigma} \Delta e_2[\phi] \, i_k \eta . \qquad (12.34)$$

The idea is again to establish that (i) the l.h.s. vanishes and (ii) the integrand on the r.h.s. is a sum of non-negative terms. The computations are more involved than before since, unlike $P[\phi]$, $e_2[\phi] \equiv \frac{1}{2} G_{AB}[\phi] (d\phi^A | d\phi^B)$ is not an algebraic expression in ϕ. The vanishing of the boundary term is, however, immediately established: Einstein's equations,

$$R_{\mu\nu} = \kappa \, G_{AB}[\phi] \, \phi^A{}_{,\mu} \, \phi^B{}_{,\nu} , \qquad (12.35)$$

imply that $R = 2\kappa e_2[\phi]$, from which we conclude that e_2 remains finite on the horizon. Since stationarity of ϕ implies $L_k e_2 = 0$, we conclude as before that the horizon integral vanishes. Asymptotic flatness requires that $dR = \mathcal{O}(r^{-3})$ and thus $*(k \wedge de_2) = \mathcal{O}(r^{-1})$. Hence, the l.h.s. of eq. (12.34) vanishes. As for (ii), we use Einstein's equations (12.35) and the matter equation $\overline{\Delta}\phi^A = 0$ (11.21) in the Bochner identity (11.26) for $\Delta e_2[\phi]$. This yields

$$\Delta e_2 = \phi_{A:\mu:\nu} \phi^{A:\mu:\nu} +$$
$$[\kappa \, G_{AB} G_{CD} - R_{ABCD}] (d\phi^A | d\phi^C)(d\phi^B | d\phi^D). \qquad (12.36)$$

(Recall that the colon and $\overline{\Delta}$ denote the covariant derivative and the Laplacian with respect to the connection $\overline{\nabla}$ along ϕ.)

Requiring that the sectional curvature of the target manifold is nonpositive, the integrand on the r.h.s. of Stoke's identity (12.34) consists of a sum of non–negative terms. Since the l.h.s. vanishes, this implies that $d\phi^A = 0$. Einstein's equations then reduce to the vacuum equations, which concludes the proof. □

In the simplest case, where ϕ is an ordinary scalar field, $\phi :$ $(M, g) \rightarrow I\!R$, the above reasoning closes the gap for free fields, which were not covered by theorem 12.8. In fact, the Bochner identity (11.27) may also be used to obtain a second proof of theorem 12.8: Using the field equations (12.27) and (12.28), the identity (11.27) yields

$$\Delta e_2 = \phi_{;\mu\nu}\phi^{;\mu\nu} + (d\phi|d\phi)\left[P_{\phi\phi} + \kappa(P + (d\phi|d\phi))\right], \qquad (12.37)$$

which provides one with the uniqueness result for both free Higgs fields and Higgs fields with *convex* potentials.

The methods presented in this section have the following short-comings: First, they cannot deal with nonconvex potentials, such as the physically relevant symmetry breaking potential $P[\phi] = (\phi^2 - \eta^2)^2$. Secondly, they yield no uniqueness results for Riemannian target manifolds with arbitrary curvature. However, as we have seen in the previous sections, neither of the problems occur for *soliton* or *spherically symmetric* black hole configurations. This clearly indicates the existence of a static no–hair theorem for self–gravitating harmonic mappings with arbitrary non–negative potentials and arbitrary Riemannian target manifolds. The problem concerning the target manifold is indeed completely removed by applying conformal techniques and the positive mass theorem (see the next section).

Regarding the potential, one might hope to get rid of the convexity restriction by taking divergence identities for Einstein's equations into account as well. The motivation for doing so is given by the Einstein–Maxwell system: Only a combination of divergence identities for both Maxwell's *and* Einstein's equations shows that the electric potential depends only on the gravitational potential - an observation which may be considered the key to the uniqueness theorem for the Reissner–Nordström metric. Following the reasoning given in section 8.3, we obtain an additional integral identity by using the general relation (2.18) (with $\omega = 0$ and

$N = -V$) and the R_{00} Einstein equation, together implying

$$d^\dagger \left(\frac{dV}{V} \right) = -\frac{2}{V} R(k, k) = 2\kappa P[\phi]. \qquad (12.38)$$

Choosing $\alpha = g(\phi, V)\frac{dV}{V}$ and taking advantage of the above relation, or choosing $\alpha = f(\phi, V)d\phi$ and using the matter equation (12.27), Stokes' identity (12.29) yields, respectively

$$8\pi \left(g_\infty M - g_H M_H \right)$$

$$= \int_\Sigma \left[g_V \frac{(dV|dV)}{V} + g_\phi \frac{(d\phi|dV)}{V} - \kappa g P \right] i_k \eta, \qquad (12.39)$$

$$- 4\pi f_\infty Q_\phi = \int_\Sigma \left[f_V (dV|d\phi) + f_\phi (d\phi|d\phi) + f P_\phi \right] i_k \eta. \qquad (12.40)$$

Here we have also used the mass formula (8.16), the definition

$$Q_\phi \equiv -\frac{1}{4\pi} \int_{S_\infty^2} *(k \wedge d\phi) \qquad (12.41)$$

of the scalar charge, and the fact that the corresponding integral over the horizon does not contribute for finite functions f. (Considering, for instance, $f = \ln(V)$, the horizon integral must be taken into account as well.) Note that for $g[\phi, V] = 1$ and $f[\phi, V] = P_\phi$, the above identities reduce to the Komar mass expression and to the identity (12.31), respectively. The hope that the above relations can be combined to yield a uniqueness result for nonconvex potentials has, to date, not been fulfilled.

12.5 The uniqueness theorem for nonrotating black holes

Using integral identities, we have seen that there exist no static black hole solutions with nontrivial harmonic mappings, provided that the target manifold has nonpositive sectional curvature. In this section we apply the conformal methods already used in the vacuum and the electrovac case, to extend this result to arbitrary Riemannian target manifolds. Since the reasoning is closely related to the vacuum case, we restrict ourselves to a brief outline of the proof, and refer to section 9.2 for details.

As in the vacuum and the electrovac cases, the horizon is assumed to be nondegenerate and is not required to be connected.

Using $R_{\mu\nu} = \kappa\, G_{AB}[\phi]\phi^A{}_{,\mu}\,\phi^B{}_{,\nu}$, the static Einstein and matter equations for self–gravitating harmonic mappings are (see eqs. (2.45)-(2.47) and (11.21)):

$$\Delta^{(g)}S = 0, \qquad (12.42)$$

$$R^{(g)} = \kappa\, G_{AB}\, g^{ij}\, \phi^A{}_{,i}\, \phi^B{}_{,j}, \qquad (12.43)$$

$$R^{(g)}_{ij} - S^{-1}\nabla^{(g)}_j\nabla^{(g)}_i S = \kappa\, G_{AB}\, \phi^A{}_{,i}\, \phi^B{}_{,j}, \qquad (12.44)$$

$$\Delta^{(g)}\phi^A + S^{-1}(dS|d\phi^A) + \Gamma^A_{BC}(\phi)(d\phi^B|d\phi^C) = 0, \qquad (12.45)$$

where, as earlier, $^{(4)}g = -S^2 dt^2 + g$.

Concerning asymptotic flatness, we require again the existence of a compact set $\mathcal{K} \in \Sigma$, such that $\Sigma - \mathcal{K}$ is diffeomorphic to $I\!\!R^3 - \overline{D}$, and such that S and the 3–dimensional metric g have the usual asymptotic expansions with respect to standard coordinates of $I\!\!R^3$ (see section 9.2). As mentioned earlier, this implies the existence of an adapted coordinate system, such that asymptotically $g = (1 + 2Mr^{-1})\delta + \mathcal{O}(r^{-2})$ and $S = 1 - Mr^{-1} + \mathcal{O}(r^{-2})$, where M denotes the total mass, and the derivatives of the terms of $\mathcal{O}(r^{-2})$ are required to be of $\mathcal{O}(r^{-3})$.

The strategy is again to show that there exists a conformal transformation of (Σ, g), such that the transformed Riemannian manifold is flat. In fact, the conformal factor which yields the desired result is the same as in the vacuum case. More precisely, we have the following proposition:

Proposition 12.10 *Consider an asymptotically flat solution of eqs. (12.42)-(12.45). Let $\hat{\Sigma}_+$, $\hat{\Sigma}_-$, $\check{\Sigma}$ and $\hat{g}_\pm$ be defined as in proposition 9.1. Then the assertions (i)-(iii) and (v) of proposition 9.1 remain true. In addition: (iv), the Ricci curvature $\hat{R}$ is non–negative.*

Proof Since S is subject to the Poisson equation $\Delta^{(g)}S = 0$, the maximum principle implies (i) as in the vacuum case. By the same reason, (ii), (iii) and (v) remain valid for the coupling to harmonic fields. Concerning the Ricci tensor $\hat{R}$, we use the transformation law (9.23), which for the conformal factor $\Omega_\pm = \frac{1}{4}(1 \pm S)^2$ yields

$$\frac{\Omega_\pm^4}{2}\,\hat{R} = \frac{\Omega_\pm^2}{2}\,R^{(g)} - \Omega_\pm\,(S \pm 1)\,\Delta^{(g)}S. \qquad (12.46)$$

Einstein's equations (12.43) now yield the desired result

$$\frac{\Omega_\pm^2}{2}\,\hat{R} \;=\; \frac{\kappa}{2}\,G_{AB}\,g^{ij}\,\phi^A{}_{,i}\,\phi^B{}_{,j} \;=\; \kappa\,e_2[\phi] \;\geq\; 0 \qquad (12.47)$$

since, for a *Riemannian* target metric, $e_2[\phi]$ is manifestly non–negative. $\qquad\qquad\qquad\qquad\qquad\qquad\qquad\qquad\qquad\Box$

The conformal 3–manifold $\hat{\Sigma}$ constructed in this way satisfies the conditions which are required to apply the positive mass theorem. As in section 9.2 we therefore conclude that $\hat{R} = 0$ and that the spatial geometry of the domain of outer communications is conformally flat. Equation (12.47) then implies that the scalar fields assume constant values and Einstein's equations reduce to the vacuum equations. The vacuum uniqueness theorem for the Schwarzschild metric then completes the proof of the following theorem:

Theorem 12.11 *Let $\phi : (M, g) \to (N, G)$ be a minimally coupled, finite, stationary harmonic mapping with Riemannian target manifold (N, G). Then the only black hole solution with regular and nonrotating event horizon and strictly stationary, asymptotically flat domain of outer communications is the Schwarzschild metric with a constant mapping ϕ_0.*

12.6 The uniqueness theorem for rotating black holes

To conclude, we extend the above uniqueness theorem to the rotating case. The circularity theorem was already established in section 12.1. The spacetime metric can therefore be written in the Papapetrou form (3.31), subject to the general identities (3.32)-(3.37).

We shall first argue that the field equations split into two decoupled sets, provided that ϕ describes a self–gravitating *harmonic* mapping. As in the vacuum case, both sets of equations can be derived from a variational principle. Since the first set is identical to the vacuum case, it uniquely determines the Ernst potential (and the associated metric functions) as the solution of a regular, 2–dimensional boundary–value problem. The second set of equations involves the matter fields and an additional metric function which is *not* determined by the Ernst potential. Using asymptotic flatness and Stokes' theorem for a suitably constructed 2–dimensional vector field, we show that these equations admit only trivial scalar field configurations. This implies that the Kerr metric is the only asymptotically flat, stationary and axisymmetric black hole solution with regular event horizon and self–gravitating harmonic mappings.

We start by recalling Einstein's equations for a harmonic mapping ϕ,

$$R_{\mu\nu} = \kappa G_{AB}[\phi]\,\phi^A_\mu\phi^B_\nu\,. \qquad (12.48)$$

Stationarity and axisymmetry of ϕ imply

$$R(m,m) = R(k,k) = R(m,k) = \mathrm{tr}_\sigma R = 0\,, \qquad (12.49)$$

where the trace is with respect to the metric σ of the 2–dimensional manifold generated by the orbits of the Killing fields k and m (see chapter 3 for the notations used in this section).

Since $R(m)$ vanishes for *harmonic* fields (see eq. (12.5)), we conclude (as in the vacuum case) that the twist $\omega \equiv \omega_m$ is closed, $d\omega = *(m \wedge R(m)) = 0$. We can therefore introduce the same Ernst potential as in the vacuum case:

$$\mathrm{E} = -X + iY\,, \quad \text{where } dY = \omega\,. \qquad (12.50)$$

Together with Einstein's equations (12.48) and the symmetry properties (12.49), the basic identities (3.32)-(3.35) for the metric functions ρ, X, A and the metric γ become

$$\Delta^{(\gamma)}\mathrm{E} + \frac{1}{\rho}(d\rho|d\mathrm{E})^{(\gamma)} + \frac{1}{X}(d\mathrm{E}|d\mathrm{E})^{(\gamma)} = 0\,, \qquad (12.51)$$

$$\Delta^{(\gamma)}\rho = 0\,, \qquad (12.52)$$

$$\frac{1}{\rho}dA = 2 *^{(\gamma)}\left(\frac{dY}{X^2}\right)\,, \qquad (12.53)$$

$$\kappa^{(\gamma)}\gamma_{ij} - \frac{1}{\rho}\nabla^{(\gamma)}_i\nabla^{(\gamma)}_j\rho = \frac{\mathrm{E}_{,i}\,\bar{\mathrm{E}}_{,j} + \mathrm{E}_{,j}\,\bar{\mathrm{E}}_{,i}}{4\,X^2} + G_{AB}\phi^A_{\,,i}\,\phi^B_{\,,j}\,, \qquad (12.54)$$

where $X = -\mathrm{Re}(\mathrm{E})$ and $Y = \mathrm{Im}(\mathrm{E})$. In addition to these equations one has the matter equations, $\Delta\phi^A + \Gamma^A_{BC}(\phi)(d\phi^A|d\phi^B) = 0$. Since the ϕ^A do not depend on t or φ, one has $(d\phi^A|d\phi^B) = X(d\phi^A|d\phi^B)^{(\gamma)}$ and $\Delta\phi^A = X[\Delta^{(\gamma)}\phi^A + (d\ln\rho|d\phi^A)^{(\gamma)}]$, and hence

$$\Delta^{(\gamma)}\phi^A + \frac{1}{\rho}(d\rho|d\phi^A)^{(\gamma)} + \Gamma^A_{BC}(d\phi^A|d\phi^B)^{(\gamma)} = 0\,. \qquad (12.55)$$

The formulae (12.51)-(12.55) represent the complete set of field equations for the functions E, ρ, ϕ^A and the 2–dimensional Riemannian metric γ. All quantities depend only on the two coordinates of the orthogonal manifold.

The crucial observation consists of the fact that no matter fields appear in the Einstein equations (12.51)-(12.53), which determine the metric of the orbit manifold. As in the vacuum case (and the Einstein–Maxwell system), ρ is a harmonic function on (Γ, γ). It is therefore again possible to choose ρ as one of the coordinates on (Γ, γ). The metric γ can then be written in the diagonal form $\gamma = e^{2h}(d\rho^2 + dz^2)$ and the spacetime metric is finally parametrized by the three functions X, A and h of the two variables ρ and z,

$$^{(4)}g = -\frac{\rho^2}{X}dt^2 + X\,(d\varphi + Adt)^2 + \frac{1}{X}e^{2h}\,(d\rho^2 + dz^2). \quad (12.56)$$

The differential equations for $h(\rho, z)$ are obtained from eq. (12.54) as in the vacuum case,

$$\frac{1}{\rho}h_{,\rho} = \frac{1}{2}(R_{\rho\rho} - R_{zz}) + \frac{1}{4X^2}\,(E_{,\rho}\bar{E}_{,\rho} - E_{,z}\bar{E}_{,z}), \quad (12.57)$$

$$\frac{1}{\rho}h_{,z} = \frac{1}{2}(R_{\rho z} + R_{z\rho}) + \frac{1}{4X^2}\,(E_{,\rho}\bar{E}_{,z} + E_{,z}\bar{E}_{,\rho}), \quad (12.58)$$

$$-\underline{\Delta}h = \frac{1}{2}(R_{\rho\rho} + R_{zz}) + \frac{1}{4X^2}\,(E_{,\rho}\bar{E}_{,\rho} + E_{,z}\bar{E}_{,z}), \quad (12.59)$$

where R_{ij} is given by eq. (12.48), $i,j \in \{\rho, z\}$ and, as earlier, underlined operators refer to the flat metric $\delta = d\rho^2 + dz^2$. The linearity of the above equations in the metric function h suggests the partition

$$h = h^{(vac)} + \kappa\,h^{(\phi)}, \quad (12.60)$$

where $h^{(vac)}$ is required to fulfil eqs. (12.57)-(12.59) with $R_{ij} = 0$. The entire set of field equations is now solved as follows:

(i) *Vacuum equations:* As in the vacuum case, one first solves the 2–dimensional boundary–value problem in the (ρ, z) plane (with fixed, flat background metric δ) for the potential E, subject to the Ernst equation

$$\frac{1}{\rho}\underline{\nabla}(\rho\,\underline{\nabla}E) + \frac{(\underline{\nabla}E|\underline{\nabla}E)}{X} = 0. \quad (12.61)$$

The metric functions A and $h^{(vac)}$ are then obtained from the Ernst potential $E = -X + iY$ by quadrature,

$$\frac{1}{\rho} A_{,\rho} = \frac{1}{X^2} Y_{,z} \,, \qquad \frac{1}{\rho} A_{,z} = -\frac{1}{X^2} Y_{,\rho} \,, \qquad (12.62)$$

$$\frac{1}{\rho} h^{(vac)}_{,\rho} = \frac{1}{4X^2} \left[E_{,\rho} \bar{E}_{,\rho} - E_{,z} \bar{E}_{,z} \right] \,,$$

$$\frac{1}{\rho} h^{(vac)}_{,z} = \frac{1}{4X^2} \left[E_{,\rho} \bar{E}_{,z} + E_{,z} \bar{E}_{,\rho} \right] \,. \qquad (12.63)$$

(ii) *Matter equations:* Like the Ernst equation, the field equations for the matter fields ϕ^A,

$$\frac{1}{\rho} \nabla(\rho \nabla \phi^A) + \Gamma^A_{BC} \left(\nabla \phi^B | \nabla \phi^C \right) = 0 \,, \qquad (12.64)$$

involve no unknown metric functions. Having solved the boundary–value problem for $\phi^A(z, \rho)$, the remaining metric function $h^{(\phi)}$ is also obtained by quadrature:

$$\frac{1}{\rho} h^{(\phi)}_{,\rho} = \frac{G_{AB}[\phi]}{2} \left[\phi^A_{,\rho} \phi^B_{,\rho} - \phi^A_{,z} \phi^B_{,z} \right] \,,$$

$$\frac{1}{\rho} h^{(\phi)}_{,z} = \frac{G_{AB}[\phi]}{2} \left[\phi^A_{,\rho} \phi^B_{,z} + \phi^A_{,z} \phi^B_{,\rho} \right] \,. \qquad (12.65)$$

Before we proceed, we recall that the Ernst equation (12.61) coincides with the integrability condition for the system (12.63) and, in addition, guarantees the consistency of these first order equations with the Poisson equation

$$-\underline{\Delta} h^{(vac)} = \frac{1}{4X^2} \left(E_{,\rho} \bar{E}_{,\rho} + E_{,z} \bar{E}_{,z} \right)$$

for the function $h^{(vac)}$ (see eq. (12.59)). In exactly the same manner, the matter equations (12.64) are identical to the integrability conditions for the system (12.65) and guarantee the consistency of these equations with the second order equation

$$-\underline{\Delta} h^{(\phi)} = \frac{G_{AB}}{2} \left[\phi^A_{,\rho} \phi^B_{,\rho} + \phi^A_{,z} \phi^B_{,z} \right] \,. \qquad (12.66)$$

As an example, we demonstrate the consistency of eqs. (12.65) and the above expression for $\underline{\Delta} h^{(\phi)}$. Differentiating $h^{(\phi)}_{,\rho}$ with respect to ρ and $h^{(\phi)}_{,z}$ with respect to z yields

$$h^{(\phi)},_{\rho\rho} + h^{(\phi)},_{zz} = \rho\,\phi^A,_{\rho} \left\{ G_{AB} \left(\phi^B,_{zz} + \phi^B,_{\rho\rho} + \frac{1}{\rho}\phi^B,_{\rho} \right) \right.$$

$$+ \left(G_{AB,C} - \frac{1}{2}G_{BC,A} \right) \phi^B,_z\,\phi^C,_z + \frac{1}{2}G_{AB,C}\,\phi^B,_{\rho}\,\phi^C,_{\rho} \left. \right\}$$

$$- \frac{G_{AB}}{2}\left(\phi^A,_{\rho}\,\phi^B,_{\rho} + \phi^A,_z\,\phi^B,_z \right) = \Delta^{(\delta)}h^{(\phi)},$$

where we have used $\Gamma_{BC|A}\phi^A,_{\rho}\,\phi^B,_z\,\phi^C,_z = (G_{AB,C} - \frac{1}{2}G_{BC,A})$
$\phi^A,_{\rho}\,\phi^B,_z\,\phi^C,_z$; $\Gamma_{BC|A}\phi^A,_{\rho}\,\phi^B,_{\rho}\,\phi^C,_{\rho} = \frac{1}{2}G_{AB,C}\,\phi^A,_{\rho}\,\phi^B,_{\rho}\,\phi^C,_{\rho}$
and the matter equation (12.64).

As in the vacuum and the electrovac cases, the second order equations for both $E(\rho,z)$ *and* $\phi^A(\rho,z)$ are the Euler–Lagrange equations for the effective Lagrangian $\mathcal{L}_{\text{eff}} = \rho\sqrt{\gamma}\kappa^{(\gamma)}$, where $\kappa^{(\gamma)}$ is the Gauss curvature of the orthogonal manifold, $\kappa^{(\gamma)} = -\Delta^{(\gamma)}h$:

$$\mathcal{L}_{\text{eff}} = \rho\left[\frac{(\nabla E|\nabla \bar{E})}{(E + \bar{E})^2} + \kappa\,\frac{G_{AB}(\nabla\phi^A|\nabla\phi^B)}{2} \right]. \qquad (12.67)$$

As usual, we introduce prolate spheroidal coordinates x and y,

$$\rho^2 = \mu^2\,(x^2 - 1)\,(1 - y^2), \qquad z = \mu\,x\,y, \qquad (12.68)$$

where μ is an arbitrary positive constant. Recall that the transformation $(\rho, z) \mapsto (x, y)$ maps the upper half–plane to the semi–strip $\mathcal{S} = \{(x,y)|x \geq 1, |y| \leq 1\}$, where the boundary $\rho = 0$ consists of the horizon, $x = 0$, and the northern and southern segments of the rotation axis, $y = 1$ and $y = -1$, respectively. In terms of x and y, the 2–dimensional metric γ assumes the form (4.20) and the effective Lagrangian for the potential $\epsilon(x,y) = (1+E)/(1-E)$ becomes

$$\mathcal{L}_{\text{eff}}^{(vac)} = \mu\,\frac{(x^2 - 1)\epsilon,_x\,\bar{\epsilon},_x + (1 - y^2)\epsilon,_y\,\bar{\epsilon},_y}{(1 - \epsilon\bar{\epsilon})^2}. \qquad (12.69)$$

Before presenting the uniqueness proof, we consider, as an example, mappings from spacetime into the pseudo–sphere PS^2,

$$\phi = (\chi, \varphi), \qquad G_{AB}[\phi] = \text{diag}\,(1\,, \sinh^2\chi).$$

Parametrizing PS^2 by complex coordinates w in the unit disc $\mathcal{D}$, $w = \tanh(\frac{\chi}{2})e^{i\varphi}$, $\mathcal{D} = \{w|w\bar{w} \leq 1\}$, we find

$$G_{AB}[\phi]\,(d\phi^A|d\phi^B) = d\chi^2 + \sinh^2\chi\,d\varphi^2 = 4\,\frac{(dw|d\bar{w})}{(1 - w\bar{w})^2}$$

for the harmonic density $e_2[\phi]$. The effective matter Lagrangian now becomes

$$\mathcal{L}_{\text{eff}}^{(\phi)} = \kappa\,\mu\,\frac{(x^2 - 1)w_{,x}\,w_{,x} + (1 - y^2)w_{,y}\,w_{,y}}{(1 - w\overline{w})^2}. \qquad (12.70)$$

Hence, if ϕ is a stationary and axisymmetric mapping from spacetime into the pseudo–sphere, then both ϵ and w satisfy exactly the same Ernst equation. Let us now prove the following theorem (Heusler 1995b):

Theorem 12.12 *Let $\phi : (M, g) \rightarrow (N, G)$ be a minimally coupled, finite, stationary harmonic mapping with Riemannian target manifold (N, G) and asymptotic behavior (12.71). Then the only black hole solution with regular event horizon and stationary, asymptotically flat domain of outer communications is the Kerr metric with a constant mapping ϕ_0.*

Proof First, the vacuum uniqueness theorem presented in chapter 10 guarantees that the quantities X, Y, A and $h^{(vac)}$, which are obtained from the Ernst potential E, are uniquely determined by the vacuum equations (12.61)-(12.63) and the boundary and regularity conditions (4.59)-(4.62). Hence, the metric of the orbit manifold is exactly the same as in the vacuum case. The only possible change in the metric must therefore be due to the function $h^{(\phi)}$. It remains to prove that $h^{(\phi)}$ vanishes identically. Once the circularity theorem is established and the basic equations are derived, this is in fact the only additional step which has to be performed in order to extend the vacuum uniqueness theorem to self–gravitating harmonic mappings.

Let us now show that the matter equations (12.64) and (12.65) admit only the trivial solution $\phi^A = \phi_0^A$, $h^{(\phi)} = 0$. We shall find that this is a consequence of asymptotic flatness, the (weak) fall–off conditions

$$\phi^A = \phi_\infty^A + \mathcal{O}(r^{-1})\,, \quad \phi^A{}_{,r} = \mathcal{O}(r^{-2})\,, \quad \phi^A{}_{,\vartheta} = \mathcal{O}(r^{-1}) \quad (12.71)$$

and the requirement that $G_{AB}[\phi]$ and the derivatives of the scalar fields with respect to r and ϑ remain finite at the boundary of the domain of outer communications. As usual, r and ϑ denote Boyer–Lindquist coordinates,

$$r = m\,(1 + p\,x)\,, \quad \cos\vartheta = y\,, \qquad (12.72)$$

in terms of which $S = \{(r, \vartheta) \,|\, r \geq r_H, \vartheta \in [0, \pi]\}$. We start by noting that eq. (12.65) for $h^{(\phi)}$ in prolate spheroidal coordinates becomes

$$
h^{(\phi)}{}_{,x} = \frac{G_{AB}}{2} \frac{1 - y^2}{x^2 - y^2} \Big\{ x(x^2 - 1)\,\phi^A{}_{,x}\,\phi^B{}_{,x}
$$
$$
- x(1 - y^2)\phi^A{}_{,y}\,\phi^B{}_{,y} - 2y(x^2 - 1)\phi^A{}_{,x}\,\phi^B{}_{,y} \Big\},
$$

$$
h^{(\phi)}{}_{,y} = \frac{G_{AB}}{2} \frac{x^2 - 1}{x^2 - y^2} \Big\{ y(x^2 - 1)\phi^A{}_{,x}\,\phi^B{}_{,x}
$$
$$
- y(1 - y^2)\phi^A{}_{,y}\,\phi^B{}_{,y} + 2x(1 - y^2)\phi^A{}_{,x}\,\phi^B{}_{,y} \Big\},
$$

from which we also obtain the asymptotic behavior of the derivative of $h^{(\phi)}$ with respect to r,

$$
h^{(\phi)}{}_{,r} = \frac{G_{AB}}{2} \Big\{ \sin^2 \vartheta \big([r + \mathcal{O}(1)]\,\phi^A{}_{,r}\,\phi^B{}_{,r}
$$
$$
- [r^{-1} + \mathcal{O}(r^{-2})]\,\phi^A{}_{,\vartheta}\,\phi^B{}_{,\vartheta} \big)
$$
$$
+ \sin(2\vartheta)\,[1 + \mathcal{O}(r^{-1})]\,\phi^A{}_{,r}\,\phi^B{}_{,\vartheta} \Big\}.
$$

The fall–off conditions (12.71) imply that all terms on the r.h.s. are maximally of $\mathcal{O}(r^{-3})$. Hence, the metric function $h^{(\phi)}(r, \vartheta)$ has the asymptotic properties

$$
\lim_{r\to\infty} r^2\, h^{(\phi)}{}_{,r} = 0\,, \qquad \lim_{r\to\infty} r\, h^{(\phi)}{}_{,\vartheta} = 0\,, \tag{12.73}
$$

where the second equation is established in a similar way.

We now apply Stokes' theorem for a suitably chosen vector field, in order to prove that eq. (12.65) has no nontrivial solutions with asymptotic behavior (12.73). Consider the vector field

$$
w = \rho\, \underline{\nabla}\, e^{-h^{(\phi)}} \tag{12.74}
$$

in the (ρ, z) plane, where $\underline{\nabla} = (\partial_\rho, \partial_z)$. Stokes's theorem for w and the domain S with counter–clockwise oriented boundary ∂S yields

$$
\oint_{\partial S} \rho e^{-h^{(\phi)}} \left(h^{(\phi)}{}_{,z}\, d\rho - h^{(\phi)}{}_{,\rho}\, dz \right) =
$$
$$
\int_S \rho e^{-h^{(\phi)}} \left[|\underline{\nabla} h^{(\phi)}|^2 - \big(\tfrac{1}{\rho} h^{(\phi)}{}_{,\rho} + \underline{\Delta} h^{(\phi)} \big) \right] d\rho dz. \tag{12.75}
$$

The key observation consists of the fact that the r.h.s. of this identity is a sum of two non–negative terms, provided that the metric

G_{AB} of the target manifold is Riemannian. This is immediately seen from eqs. (12.65) and (12.66), which together yield

$$- (h^{(\phi)}{}_{,\rho} + \rho\underline{\Delta}h^{(\phi)}) = \rho\, G_{AB}[\phi]\, \phi^A{}_{,z}\, \phi^B{}_{,z} \geq 0. \qquad (12.76)$$

Our final task is to show that the boundary integral on the l.h.s. of eq. (12.75) vanishes, implying that $h^{(\phi)}$ is constant. In terms of prolate spheroidal coordinates, we must establish that $\lim_{R\to\infty} I_R = 0$, where

$$I_R = \mu \oint_{\partial S_R} e^{-h^{(\phi)}} \left[(1 - y^2)h^{(\phi)}{}_{,y}\, dx - (x^2 - 1)h_{,x}\, dy \right] , \qquad (12.77)$$

and where the oriented boundary ∂S_R is the rectangle $\gamma_R^1 = \{y = 1, x = R...1\}$, $\gamma^2 = \{x = 1, y = 1... - 1\}$, $\gamma_R^3 = \{y = -1, x = 1...R\}$ and $\gamma_R^4 = \{x = R, y = -1...1\}$. The finiteness of the Ricci scalar and the regularity of the derivatives of $h^{(\phi)}$ with respect to Boyer–Lindquist coordinates imply that $h^{(\phi)}{}_{,x}$, $h^{(\phi)}{}_{,y}$ and $\exp(-h^{(\phi)})$ remain finite along γ_R^1, γ^2 and γ_R^3. Hence, both integrals in eq. (12.77) vanish on these parts of the boundary. It remains to compute the contribution from the integration along γ_R^4 as $R \to \infty$, that is

$$\lim_{R\to\infty} I_R = -\mu \lim_{R\to\infty} \int_{\gamma_R^4} e^{-h^{(\phi)}} (x^2 - 1)\, h_{,x}\, dy$$

$$= - \int_0^\pi \lim_{r\to\infty} (e^{-h^{(\phi)}} r^2 h_{,r})\, \sin\vartheta\, d\vartheta. \qquad (12.78)$$

The asymptotic behavior (12.73) of $r^2 h_{,r}$ and the fact that $h^{(\phi)}$ is bounded imply that this integral vanishes as well. The l.h.s. of eq. (12.75) is therefore zero, which enables us to conclude that both integrands on the r.h.s., $|\underline{\nabla}h^{(\phi)}|^2$ and $-[h^{(\phi)}{}_{,\rho} + \rho\underline{\Delta}h^{(\phi)}]$, must vanish. Hence, $h^{(\phi)}$ is constant in all of the domain of outer communications. Since $\lim_{r\to\infty} h^{(\phi)} = 0$, we finally have

$$h^{(\phi)} = 0. \qquad (12.79)$$

This proves that ϕ is constant and establishes the uniqueness of the Kerr metric amongst the black hole solutions with *circular* domain of outer communications.

Together with the strong rigidity theorem, the staticity and circularity theorems and the static uniqueness theorem 12.11, we finally obtain the uniqueness of the Kerr metric amongst *all stationary* black hole solutions of self-gravitating harmonic mappings with Riemannian target manifolds. □

References

Adkins G.S., Nappi C.R. & Witten E. (1983), Static Properties of
Nucleons in the Skyrme Model. *Nucl. Phys. B* **228**: 552–556.

Arnowitt R., Deser S. & Misner C.W. (1962), The Dynamics of
General Relativity. In: *Gravitation: An Introduction to Current
Research*, ed. L. Witten. New York, Plenum.

Ashtekar A. (1980), Asymptotic Structure of the Gravitational Field at
Spatial Infinity. In: *General Relativity and Gravitation 2*, ed. A.
Held. New York, Plenum.

Ashtekar A. (1984), On the Boundary Conditions for Gravitational and
Gauge Fields at Spatial Infinity. In: *Asymptotic Behavior of Mass
and Spacetime Geometry*. Lecture Notes in Physics **202**, ed. F.J.
Flaherty. New York, Springer.

Ashtekar A. & Hansen R.O. (1978), A Unified Treatment of Null and
Spatial Infinity in General Relativity. *J. Math. Phys.* **19**: 1542–1566.

Ashtekar A. & Magnon–Ashtekar A. (1979), On Conserved Quantities
in General Relativity. *J. Math. Phys.* **20**: 793–800.

Bardeen J.M., Carter B. & Hawking S.W. (1973), The Four Laws of
Black Hole Mechanics. *Commun. Math. Phys.* **31**: 161–170.

Bartnik R. (1984), The Existence of Maximal Surfaces in
Asymptotically Flat Spacetimes. *Commun. Math. Phys.* **94**:
155–175.

Bartnik R. (1986), The Mass of an Asymptotically Flat Manifold.
Commun. Pure Appl. Math. **39**: 661–693.

Bartnik R. & McKinnon J. (1988), Particle–Like Solutions of the
Einstein–Yang–Mills Equations. *Phys. Rev. Lett.* **61**: 141–144.

Bechmann O. & Lechtenfeld O. (1995), Exact Black–Hole Solution with
Self–Interacting Scalar Field. *Class. Quantum Grav.* **12**: 1473–1481.

Behnke H. & Sommer F. (1976), *Theorie der Analytischen Funktionen
einer Komplexen Veränderlichen*. Berlin, Springer.

Beig R. (1978), Arnowitt–Deser–Misner Energy and g_{00}. *Phys. Lett.
A* **69**: 153–155.

Beig R. & Simon W. (1980a), Proof of a Multipole Conjecture Due to
Geroch. *Commun. Math. Phys.* **78**: 75–82.

Beig R. & Simon W. (1980b), The Stationary Gravitational Field near Spacelike Infinity. *Gen. Rel. Grav.* **12**: 439–451.

Beig R. & Simon W. (1981), On the Multipole Expansion for Stationary Space–Times. *Proc. R. Soc. London Ser. A* **376**: 333–341.

Bekenstein J.D. (1972), Nonexistence of Baryon Number for Static Black Holes. *Phys. Rev. D* **5**: 1239–1246.

Bekenstein J.D. (1973), Extraction of Energy and Charge from a Black Hole. *Phys. Rev. D* **7**: 949–953.

Bekenstein J.D. (1974a), Generalized Second Law of Thermodynamics in Black Hole Physics. *Phys. Rev. D* **9**: 3292–3300.

Bekenstein J.D. (1974b), Exact Solutions of Einstein–Conformal Scalar Equations. *Ann. Phys. (NY)* **82**: 535–547.

Bekenstein J.D. (1975), Black Holes with Scalar Charge. *Ann. Phys. (NY)* **91**: 75–82.

Birkhoff G.D. (1923), *Relativity and Modern Physics*. Cambridge MA, Harvard Univ. Press.

Bishop R.L. & Crittendon R.J. (1964), *Geometry of Manifolds*. New York, Academic Press.

Bizon P. (1990), Colored Black Holes. *Phys. Rev. Lett.* **64**: 2844–2847.

Bochner S. (1940), Harmonic Surfaces in Riemannian Metric. *Trans. Am. Math. Soc.* **47**: 146–149.

Bogomol'nyi E.B. (1976), The Stability of Classical Solutions. *Sov. J. Nucl. Phys.* **24**: 449–454.

Bondi H. (1960), Gravitational Waves in General Relativity. *Nature* **186**: 535.

Bondi H., van de Burg M.G.J. & Metzner A.W.K. (1962), Gravitational Waves in General Relativity. VII. Waves for Axisymmetric Isolated Systems. *Proc. R. Soc. London Ser. A* **269**: 21–52.

Boothby W.M. (1975), *An Introduction to Differentiable Manifolds and Riemannian Geometry*. New York, Academic Press.

Boyer R.H. & Lindquist R.W. (1967), Maximal Analytic Extension of the Kerr Metric. *J. Math. Phys.* **8**: 265–281.

Breitenlohner P., Maison D. & Gibbons G.W. (1988), 4–Dimensional Black Holes from Kaluza–Klein Theories. *Commun. Math. Phys.* **120**: 295–334.

Breitenlohner P., Forgács P. & Maison D. (1992), Gravitating Monopole Solutions. *Nucl. Phys. B* **383**: 357–376.

Brill D.R. (1964), Electromagnetic Fields in Homogeneous, Nonstatic Universe. *Phys. Rev.* **133**: B845–B848.

Brodbeck O. & Straumann N. (1993), A Generalized Birkhoff Theorem for the Einstein–Yang–Mills System. *J. Math. Phys.* **34**: 2412–2423.

Brodbeck O. & Straumann N. (1994), Self–Gravitating Yang–Mills
  Solutions and their Chern–Simons Numbers. *J. Math. Phys.* **35**:
  899–919.

Brodbeck O., Heusler M. & Straumann N. (1996), Pulsation of
  Spherically Symmetric Systems in General Relativity. *Phys. Rev.*
  D **53**: 754–765.

Brown J.D. & York J.W. (1993a), Quasilocal Energy and Conserved
  Charges Derived from the Gravitational Action. *Phys. Rev. D* **47**:
  1407–1419.

Brown J.D. & York J.W. (1993b), Microcanonical Functional Integral
  for the Gravitational Field. *Phys. Rev. D* **47**: 1420–1431.

Bunting G.L. (1983), Proof of the Uniqueness Conjecture for Black
  Holes. *PhD Thesis*, Univ. of New England, Armidale, N.S.W.

Bunting G.L. & Masood-ul-Alam A.K.M. (1987), Nonexistence of
  Multiple Black Holes in Asymptotically Euclidean Static Vacuum
  Space–Times. *Gen. Rel. Grav.* **19**: 147–154.

Callen C.G., Coleman S. & Jackiw R. (1970), A New Improved
  Energy–Momentum Tensor. *Ann. Phys. (NY)* **59**: 42–73.

Carter B. (1968), Global Structure of the Kerr Family of Gravitational
  Fields. *Phys. Rev.* **174**: 1559–1571.

Carter B. (1969), Killing Horizons and Orthogonally Transitive Groups
  in Space–Time. *J. Math. Phys.* **10**: 70–81.

Carter B. (1970), The Commutation Property of a Stationary,
  Axisymmetric System. *Commun. Math. Phys.* **17**: 233–238.

Carter B. (1971), Axisymmetric Black Hole has only Two Degrees of
  Freedom. *Phys. Rev. Lett.* **26**: 331–332.

Carter B. (1973a), Black Hole Equilibrium States. In: *Black Holes*, eds.
  C. DeWitt & B.S. DeWitt. New York, Gordon & Breach.

Carter B. (1973b), Rigidity of a Black Hole. *Nature (Phys. Sci.)* **238**:
  71–72.

Carter B. (1973c), Elastic Perturbation Theory in General Relativity
  and a Variational Principle for a Rotating Solid Star. *Commun.
  Math. Phys.* **30**: 261–268.

Carter B. (1979), The General Theory of the Mechanical,
  Electromagnetic and Thermodynamic Properties of Black Holes. In:
  General Relativity, an Einstein Centenary Survey, eds. S.W.
  Hawking & W. Israel. Cambridge Univ. Press.

Carter B. (1985), Bunting Identity and Mazur Identity for Non–Linear
  Elliptic Systems Including the Black Hole Equilibrium Problem.
  Commun. Math. Phys. **99**: 563–591.

Carter B. (1987), Mathematical Foundations of the Theory of Relativistic Stellar and Black Hole Configurations. In: *Gravitation in Astrophysics*, eds. B. Carter & J.B. Hartle. New York, Plenum.

Chandrasekhar S. (1931a), The Maximum Mass of Ideal White Dwarfs. *Astrophys. J.* **74**: 81–82.

Chandrasekhar S. (1931b), Highly Collapsed Configurations of Stellar Mass. *Mon. Not. R. Astron. Soc.* **91**: 456–466.

Chandrasekhar S. (1939), *An Introduction to the Study of Stellar Structure*. Univ. of Chicago Press.

Chandrasekhar S. (1983), *The Mathematical Theory of Black Holes*. Oxford, Clarendon Press.

Chandrasekhar S. (1989), How One May Explore the Physical Content of the General Theory of Relativity. In: *Proceedings of the Gibbs Symposium*. Yale University, The American Math. Soc.

Chandrasekhar S. (1991), *The Mathematical Theory of Black Holes and of Colliding Plane Waves*. Selected Papers, Vol 6. Univ. of Chicago Press.

Chen B. (1981), *Geometry of Submanifolds and its Applications*. Science Univ. of Tokyo.

Choquet–Bruhat Y., DeWitt–Morette C. & Dillard–Bleick M. (1982), *Analysis, Manifolds and Physics*. New York, North–Holland Publishing.

Christodoulou D. & Ó Murchadha N. (1981), The Boost Problem in General Relativity. *Commun. Math. Phys.* **80**: 271–300.

Chruściel P.T. (1987), On Angular Momentum at Spatial Infinity. *Class. Quantum Grav.* **4**: L205–L210.

Chruściel P.T. (1989a), On the Structure of Spatial Infinity. I. The Geroch Structure. *J. Math. Phys.* **30**: 2090–2093.

Chruściel P.T. (1989b), On the Structure of Spatial Infinity. II. Geodesically Regular Ashtekar–Hansen Structures. *J. Math. Phys.* **30**: 2094–2100.

Chruściel P.T. (1991), *On Uniqueness in the Large of Solutions of Einstein Equations ("Strong Cosmic Censorship")*. Canberra, Australian Univ. Press.

Chruściel P.T. (1993), On Completeness of Orbits of Killing Vector Fields. *Class. Quantum Grav.* **10**: 2091–2101.

Chruściel P.T. (1994), "No–Hair" Theorems: Folklore, Conjectures, Results. In: *Differential Geometry and Mathematical Physics*, eds. J. Beem & K.L. Duggal. Am. Math. Soc., Providence.

Chruściel P.T. & Nadirashvili N.S. (1995), All Electrovac Majumdar–Papapetrou Spacetimes with Non–Singular Black Holes. *Class. Quantum Grav.* **12**: L17–L23.

Chruściel P.T. & Wald R.M. (1994a), Maximal Hypersurfaces in Stationary Asymptotically Flat Space–Times. *Commun. Math. Phys.* **163**: 561–604.

Chruściel P.T. & Wald R.M. (1994b), On the Topology of Stationary Black Holes. *Class. Quantum Grav.* **11**: L147–L152.

Clarke C.J.S. (1975), Singularities in Globally Hyperbolic Space–Time. *Commun. Math. Phys.* **41**: 65–78.

Clarke C.J.S. (1993), *The Analysis of Space–Time Singularities.* Cambridge Lecture Notes in Physics, Cambridge Univ. Press.

Coleman S., Park S., Neveu A. & Sommerfeld C.M. (1977), Can One Dent a Dyon? *Phys. Rev. D* **15**: 544–545.

Collinson C.D. (1970), Curvature Collineations in Empty Spacetime. *J. Math. Phys.* **11**: 818–819.

Davies P.C.W. (1978), Thermodynamics of Black Holes. *Rep. Prog. Phys.* **41**: 1313–1355.

DeFelice F. & Clarke C.J.S. (1990), *Relativity on Curved Manifolds.* Cambridge Univ. Press.

Detweiler S. (1982), *Black Holes: Selected Reprints.* Stony Brook, NY, Am. Assoc. of Phys. Teachers.

Droz S., Heusler M. & Straumann N. (1991), New Black Hole Solutions with Hair. *Phys. Lett. B* **268**: 371–376.

Duff M.J. & Isham C.J. (1977), Form–Factor Interpretation of Kink Solutions to the Non–Linear σ–Model. *Phys. Rev. D* **16**: 3047–3059.

Eells J. & Lemaire L. (1978), A Report on Harmonic Mappings. *Bull. London Math. Soc.* **10**: 1–68.

Eells J. & Lemaire L. (1988), Another Report on Harmonic Mappings. *Bull. London Math. Soc.* **20**: 385–524.

Eells J. & Samson H.J. (1964), Harmonic Mappings of Riemannian Manifolds. *Am. J. Math.* **86**: 109–160.

Eichenherr H. & Forger M. (1980), More about Non–Linear Sigma–Models on Symmetric Spaces. *Nucl. Phys. B* **164**: 528–535.

Einstein A. (1915a), Zur Allgemeinen Relativitätstheorie. *Preuss. Akad. Wiss. Berlin, Sitz.ber. II* : 778–786, 799–801.

Einstein A. (1915b), Die Feldgleichungen der Gravitation. *Preuss. Akad. Wiss. Berlin, Sitz.ber. II* : 844–847.

Eisenhart L.P. (1949), *Riemannian Geometry.* Princeton Univ. Press.

Ernst F.J. (1968a), New Formulation of the Axially Symmetric Gravitational Field Problem. *Phys. Rev.* **167**: 1175–1178.

Ernst F.J. (1968b), New Formulation of the Axially Symmetric Gravitational Field Problem. II. *Phys. Rev.* **168**: 1415–1417.

Esteban M.J. (1986), A Direct Variational Approach to Skyrme's Model. *Commun. Math. Phys.* **105**: 571–591.

Fischer A.E. & Marsden J.E. (1972), The Einstein Evolution Equations as a First–Order Symmetric Hyperbolic Quasilinear System. *Commun. Math. Phys.* **28**: 1–38.

Fischer A.E. & Marsden J.E. (1976), A New Hamiltonian Structure for the Dynamics of General Relativity. *Gen. Rel. Grav.* **12**: 915–920.

Fischer A.E. & Marsden J.E. (1979), The Initial Value Problem and the Dynamical Formulation of General Relativity. In: *General Relativity, an Einstein Centenary Survey*, eds. S.W. Hawking & W. Israel. Cambridge Univ. Press.

Forgács P. & Manton N.S. (1980), Space–Time Symmetries in Gauge Theories. *Commun. Math. Phys.* **72**: 15–35.

Friedman J.L., Schleich K. & Witt D.M. (1993), Topological Censorship. *Phys. Rev. Lett.* **71**: 1486–1489.

Fuller F.B. (1954), Harmonic Mappings. *Proc. Natl. Acad. Sci.* **40**: 987–991.

Galloway G. (1993), On the Topology of Black Holes. *Commun. Math. Phys.* **151**: 53–66.

Galloway G. (1994), Least Area Tori, Black Holes and Topological Censorship. In: *Differential Geometry and Mathematical Physics*, eds. J. Beem & K.L. Duggal. Providence, Am. Math. Soc.

Gal'tsov D.V. & Xanthopoulos B.C. (1992), A Generating Technique for Einstein Gravity Conformally Coupled to a Scalar Field with Higgs Potential. *J. Math. Phys.* **33**: 273–277.

Gell–Mann M. & Levy M. (1960), The Axial Vector Current in Beta Decay. *Nuovo Cimento* **16**: 705–726.

Geroch R.P. (1970a), The Domain of Dependence. *J. Math. Phys.* **11**: 343–348.

Geroch R.P. (1970b), Multipole Moments. II. Curved Space. *J. Math. Phys.* **11**: 2580–2588.

Geroch R.P. (1971), A Method for Generating Solutions of Einstein's Equations. *J. Math. Phys.* **12**: 918–924.

Geroch R.P. (1972a), A Method for Generating New Solutions of Einstein's Equations. *J. Math. Phys.* **13**: 394–404.

Geroch R.P. (1972b), Structure of the Gravitational Field at Spatial Infinity. *J. Math. Phys.* **13**: 956–968.

Geroch R.P. (1973), Energy Extraction. In: *Sixth Texas Symposium on Relativistic Astrophysics. Ann. N.Y. Acad. Sci.* **224**: 108–117.

Geroch R.P. (1976), Asymptotic Structure of Space–Time. In: *Asymptotic Structure of Space–Time*, eds. P. Esposito & L. Witten. New York, Plenum.

Geroch R.P. & Horowitz. (1979), Global Structure of Spacetimes. In:
General Relativity, an Einstein Centenary Survey,
eds. S.W. Hawking & W. Israel. Cambridge Univ. Press.

Gibbons G.W. (1991), Self–Gravitating Magnetic Monopoles, Global
Monopoles and Black Holes. In: *The Physical Universe: The
Interface Between Cosmology, Astrophysics and Particle Physics.*
Lecture Notes in Physics **383**, ed. J.D. Barrow. New York, Springer.

Gibbons G.W. & Hawking S.W. (1977), Action Integrals and Partition
Functions in Quantum Gravity. *Phys. Rev. D* **15**: 2752–2756.

Gibbons G.W. & Hull C.M. (1982), A Bogomol'nyi Bound for General
Relativity and Solitons in $N = 2$ Supergravity. *Phys. Lett.* **109**:
190–194.

Gibbons G.W., Hawking S.W., Horowitz G.T. & Perry M.J. (1983),
Positive Mass Theorems for Black Holes. *Commun. Math. Phys.* **88**:
295–308.

Greene B.R., Mathur S.D. & O'Neill C.M. (1993), Eluding the
No–Hair Conjecture: Black Holes in Spontaneously Broken Gauge
Theories. *Phys. Rev. D* **47**: 2242–2259.

Hajicek P. (1973), General Theory of Vacuum Ergospheres. *Phys. Rev.
D* **7**: 2311–2316.

Hajicek P. (1975), Stationary Electrovac Spacetimes with Bifurcate
Horizon. *J. Math. Phys.* **16**: 518–527.

Harnad J., Shnider S. & Vinet L. (1980), Group Actions on Principal
Bundles and Invariance Conditions for Gauge Fields. *J. Math. Phys.*
21: 2719–2724.

Harrison B.K., Thorne K.S., Wakano M. & Wheeler J.A. (1965),
Gravitation Theory and Gravitational Collapse. Univ. of Chicago
Press.

Hartle J.B. & Hawking S.W. (1972), Solutions of the Einstein–Maxwell
Equations with Many Black Holes. *Comm. Math. Phys.* **26**: 87–101.

Hawking S.W. (1972), Black Holes in General Relativity. *Commun.
Math. Phys.* **25**: 152–166.

Hawking S.W. (1973), The Event Horizon. In: *Black Holes,*
eds. C. DeWitt & B.S. DeWitt. New York, Gordon and Breach.

Hawking S.W. (1975), Particle Creation by Black Holes. *Commun.
Math. Phys.* **43**: 199–220.

Hawking S.W. & Ellis G.F.R. (1973), *The Large Scale Structure of
Space Time.* Cambridge Univ. Press.

Hawking S.W. & Penrose R. (1970), The Singularities of Gravitational
Collapse and Cosmology. *Proc. R. Soc. London Ser. A* **314**:
529–548.

Hayward S.A. (1993), Dual–Null Dynamics of the Einstein Field. *Class. Quantum Grav.* **10**: 779–790.

Hayward S.A. (1994a), General Laws of Black Hole Dynamics. *Phys. Rev. D* **49**: 6467–6474.

Hayward S.A. (1994b), Quasilocal Gravitational Energy. *Phys. Rev. D* **49**: 831–839.

Hayward S.A. (1994c), Spin Coefficient Form of the New Laws of Black Hole Dynamics. *Class. Quantum Grav.* **11**: 3025–3035.

Hayward S.A. (1994d), Gravitational Energy in Spherical Space–Times. *Kyoto Univ. Preprint*, gr-qc/9408002.

Helgason S. (1962), *Differential Geometry and Symmetric Spaces.* New York, Academic Press.

Heusler M. (1992), A No–Hair Theorem for Self–Gravitating Nonlinear Sigma Models. *J. Math. Phys.* **33**: 3497–3502.

Heusler M. (1993), Staticity and Uniqueness of Multiple Black Hole Solutions of Sigma Models. *Class. Quantum Grav.* **10**: 791–799.

Heusler M. (1994), On the Uniqueness of the Reissner–Nordström Solution with Electric and Magnetic Charge. *Class. Quantum Grav.* **11**: L49–L53.

Heusler M. (1995a), A Mass Bound for Spherically Symmetric Black Hole Spacetimes. *Class. Quantum Grav.* **12**: 779–789.

Heusler M. (1995b), The Uniqueness Theorem for Rotating Black Hole Solutions of Self–gravitating Harmonic Mappings. *Class. Quantum Grav.* **12**: 2021–2035.

Heusler M. (1995c), On the Uniqueness of the Papapetrou–Majumdar Metric. *Enrico Fermi Inst. Preprint*, Univ. of Chicago; to appear in *Class. Quantum Grav.*

Heusler M. & Straumann N. (1992), Scaling Arguments for the Existence of Static, Spherically Symmetric Solutions of Self–Gravitating Systems. *Class. Quantum Grav.* **9**: 2177–2189.

Heusler M. & Straumann N. (1993a), The First Law of Black Hole Physics for a Class of Non–Linear Matter Models. *Class. Quantum Grav.* **10**: 1299–1321.

Heusler M. & Straumann N. (1993b), Mass Variation Formulae for Einstein–Yang–Mills–Higgs and Einstein–Dilaton Black Holes. *Phys. Lett. B* **315**: 55–66.

Heusler M., Droz S. & Straumann N. (1991), Stability Analysis of Self–Gravitating Skyrmions. *Phys. Lett. B* **271**: 61–67.

Heusler M., Droz S. & Straumann N. (1992), Linear Stability of Einstein–Skyrme Black Holes. *Phys. Lett. B* **285**: 21–26.

Heusler M., Straumann N. & Zhou Z-H. (1993), Self–Gravitating
  Solutions of the Skyrme Model and their Stability. *Helv. Phys. Acta*
  66: 614–632.

Hodge W.V.D. (1959), *The Theory and Applications of Harmonic
  Integrals*. Cambridge Univ. Press.

Hoenselars C., Kinnersley W. & Xanthopoulos B.C. (1979),
  Symmetries of the Stationary Einstein–Maxwell Equations. VI.
  Transformations which Generate Asymptotically Flat Spacetimes
  with Arbitrary Multipole Moments. *J. Math. Phys.* **20**: 2530–2536.

Horowitz G. T. & Perry M.J. (1982), Gravitational Energy Cannot
  Become Negative. *Phys. Rev. Lett.* **48**: 371–374.

Hungerbühler N. (1994), *P*-Harmonic Flow. *PhD Thesis* 10740,
  ETH Zürich.

Israel W. (1967), Event Horizons in Static Vacuum Space–Times.
  Phys. Rev. **164**: 1776–1779.

Israel W. (1968), Event Horizons in Static Electrovac Space–Times.
  Commun. Math. Phys. **8**: 245–260.

Israel W. (1986), Third Law of Black Hole Dynamics: A Formulation
  and Proof. *Phys. Rev. Lett.* **57**: 397–399.

Israel W. (1987), Dark Stars: The Evolution of an Idea. In: *300 Years
  of Gravitation*, eds. S.W. Hawking & W. Israel. Cambridge Univ.
  Press.

Israel W. & Wilson G.A. (1972), A Class of Stationary
  Electromagnetic Vacuum Fields. *J. Math. Phys.* **13**: 865–867.

Iyer V. & Wald R.M. (1994), Some Properties of Noether Charge and a
  Proposal for Dynamical Black Hole Entropy. *Phys. Rev. D* **50**:
  846–864.

Jackiw R. & Manton N.S. (1980), Symmetries and Conservation Laws
  in Gauge Theories. *Ann. Phys.* **127**: 257–273.

Jackson A., Jackson A.D., Goldhaber A.S., Brown G.E. & Castillejo
  L.C. (1985), A Modified Skyrmion. *Phys. Lett. B* **154**: 101–106.

Jacobson J.A. & Kang G. (1993), Conformal Invariance of Black Hole
  Temperature. *Class. Quantum Grav.* **10**: L201–L206.

Jacobson J.A., Kang G. & Myers R.C. (1994), On Black Hole Entropy.
  Phys. Rev. D **49**: 6587–6598.

Jacobson T. & Venkataramani S. (1995), Topology of Event Horizon
  and Topological Censorship. *Class. Quantum Grav.* **12**: 1055–1061.

Jang P.S. (1978), On the Positivity of Energy in General Relativity.
  J. Math. Phys. **19**: 1152–1155.

Jang P.S. & Wald R.M. (1977), The Positive Energy Conjecture and
  the Cosmic Censor Hypothesis. *J. Math. Phys.* **18**: 41–44.

Janis A.I., Robinson D.C. & Winicour J. (1969), Comments on Einstein Scalar Solutions. *Phys. Rev.* **186**: 1729–1731.

Kay B.S. & Wald R.M. (1991), Theorems on the Uniqueness and Thermal Properties of Stationary, Nonsingular, Quasifree States on Spacetimes with a Bifurcate Horizon. *Phys. Rep.* **207**: 49–136.

Kennefick D. & Ó Murchadha N. (1995), Weakly Decaying Asymptotically Flat Static and Stationary Solutions to the Einstein Equations. *Class. Quantum Grav.* **12**: 149–158.

Kerr R.P. (1963), Gravitational Field of a Spinning Mass as an Example of Algebraically Special Metrics. *Phys. Rev. Lett.* **11**: 237–238.

Kerr R.P. & Schild A. (1965), A New Class of Vacuum Solutions of the Einstein Field Equations. In: *Proceedings of the Galileo Galilei Centenary Meeting on General Relativity, Problems of Energy and Gravitational Waves*, ed. G. Barbera. Comitato Nazionale per le Manifestazione Celebrative, Florence, Italy.

Kinnersley W. (1973), Generation of Stationary Einstein–Maxwell Fields. *J. Math. Phys.* **14**: 651–653.

Kinnersley W. (1977), Symmetries of the Stationary Einstein–Maxwell Field Equations. I. *J. Math. Phys.* **18**: 1529–1537.

Kinnersley W. & Chitre D.M. (1977), Symmetries of the Stationary Einstein–Maxwell Field Equations. II. *J. Math. Phys.* **18**: 1538–1542.

Kinnersley W. & Chitre D.M. (1978a), Symmetries of the Stationary Einstein–Maxwell Field Equations. III. *J. Math. Phys.* **19**: 1926–1931.

Kinnersley W. & Chitre D.M. (1978b), Symmetries of the Stationary Einstein–Maxwell Field Equations. IV. Transformations which Preserve Asymptotic Flatness. *J. Math. Phys.* **19**: 2037–2042.

Kobayashi S. & Nomizu K. (1969), *Foundations of Differential Geometry*. New York, Interscience Publishers.

Komar A. (1959), Covariant Conservation Laws in General Relativity. *Phys. Rev.* **113**: 934–936.

Komar A. (1962), Asymptotic Covariant Conservation Laws for Gravitational Radiation *Phys. Rev.* **127**: 1411–1418.

Kramer D., Stephani H., MacCallum M. & Herlt E. (1980), *Exact Solutions of Einstein's Equations*. Cambridge Univ. Press.

Kruskal M.D. (1960), Maximal Extension of Schwarzschild Metric. *Phys. Rev.* **119**: 1743–1745.

Künzle H.P. (1971), On the Spherical Symmetry of a Static Perfect Fluid. *Commun. Math. Phys.* **20**: 85–100.

Künzle H.P. & Masood–ul–Alam A.K.M. (1990), Spherically Symmetric Static SU(2) Einstein–Yang–Mills Fields. *J. Math. Phys.* **31**: 928–935.

Kundt W. & Trümper M. (1966), *Ann. Physik.* **192**: 414–418.

Landau L. & Lifshitz E. (1971), *The Classical Theory of Fields*. London, Addison Wesley.

Laplace P.S. (1796), *Exposition du Système du Monde*. Paris, J.B.M. Duprat.

Lavrelashvili G. & Maison D. (1993), Regular and Black Hole Solutions of Einstein–Yang–Mills–Dilaton Theory. *Nucl. Phys. B* **410**: 407–422.

Li Y. & Tian G. (1992), Regularity of Harmonic Maps with Prescribed Singularities. *Commun. Math. Phys.* **149**: 1–30.

Lichnerowicz A. (1955), *Théories Relativistes de la Gravitation et de l'Electromagnétisme*. Paris, Masson.

Lindblom L. (1980), The Properties of Static General Relativistic Stellar Models. *J. Math. Phys.* **21**: 1455–1459.

Lindblom L. & Masood–ul–Alam A.K.M. (1994), On the Spherical Symmetry of Static Stellar Models. *Commun. Math. Phys.* **162**: 123–145.

Luckock H. & Moss I. (1986), Black Holes have Skyrme Hair. *Phys. Lett. B* **176**: 341–345.

Ludvigson M. & Vickers J.A.G. (1983), An Inequality Relating Total Mass and the Area of a Trapped Surface in General Relativity. *J. Phys. A: Math. Gen.* **16**: 3349–3353.

Majumdar S.D. (1947), A Class of Exact Solutions of Einstein's Field Equations. *Phys. Rev.* **72**: 390–398.

Manton N.S. (1987), Geometry of Skyrmions. *Commun. Math. Phys.* **111**: 469–478.

Manton N.S. & Ruback P.J. (1986), Skyrmions in Flat Space and Curved Space. *Phys. Lett. B* **181**: 137–140.

Masood–ul–Alam A.K.M. (1987), The Topology of Asymptotically Euclidean Static Perfect Fluid Space–Time. *Commun. Math. Phys.* **108**: 193–211.

Masood–ul–Alam A.K.M. (1992), Uniqueness Proof of Static Black Holes Revisited. *Class. Quantum Grav.* **9**: L53–L55.

Masood–ul–Alam A.K.M. (1993), Uniqueness of a Static Charged Dilaton Black Hole. *Class. Quantum Grav.* **10**: 2649–2656.

Matsushima Y. (1972), *Differentiable Manifolds*. New York, Marcel Dekker.

Mazur P.O. (1982), Proof of Uniqueness of the Kerr–Newman Black Hole Solution. *J. Phys. A: Math. Gen.* **15**: 3173–3180.

Mazur P.O. (1984a), Black Hole Uniqueness from a Hidden Symmetry of Einstein's Gravity. *Gen. Rel. Grav.* **16**: 211–215.

Mazur P.O. (1984b), A Global Identity for Nonlinear Sigma–Models. *Phys. Lett. A* **100**: 341–344.

Michell J. (1784), On the Means of Discovering the Distance... *Phil. Trans. R. Soc. (London)* **74**: 35–57; see Detweiler (1982).

Milnor J. (1963), *Morse Theory.* Princeton Univ. Press.

Misner C.W. (1965), The Flatter Regions of Newman, Unti, and Tamburino's Generalized Schwarzschild Space. *J. Math. Phys.* **4**: 924–937.

Misner C.W. (1978), Harmonic Maps as Models for Physical Theories. *Phys. Rev. D* **18**: 4510–4524.

Misner C.W. & Sharp D.H. (1964), Relativistic Equations for Adiabatic, Spherically Symmetric Gravitational Collapse. *Phys. Rev. (Sect. B)* **136**: 571–576.

Müller zum Hagen H, Robinson D.C. & Seifert H.J. (1973), Black Hole in Static Vacuum Space–Times. *Gen. Rel. Grav.* **4**: 53–78.

Müller zum Hagen H, Robinson D.C. & Seifert H.J. (1974), Black Holes in Static Electrovac Space–Times. *Gen. Rel. Grav.* **5**: 61–72.

Neugebauer G. & Kramer D. (1969), Eine Methode zur Konstruktion stationärer Einstein–Maxwell–Felder. *Ann. Physik.* **24**: 62–71.

Newman E.T. & Penrose R. (1962), An Approach to Gravitational Radiation by a Tetrad of Spin Coefficients. *J. Math. Phys.* **3**: 566–578. (erratum **4**: 998.)

Newman E.T. & Tod K.P. (1980), Asymptotically Flat Space–Times. In: *General Relativity and Gravitation 2*, ed. A. Held. New York, Plenum.

Newman E.T., Tamburino L. & Unti T. (1963), Empty–Space Generalization of the Schwarzschild Metric. *J. Math. Phys.* **4**: 915–923.

Newman E.T., Couch E., Chinnapared E., Exton K., Prakash A. & Torrence R. (1965), Metric of Rotating, Charged Mass. *J. Math. Phys.* **6**: 918–919.

Nordström G. (1918), On the Energy of the Gravitational Field in Einstein's Theory. *Proc. Kon. Ned. Akad. Wet.* **20**: 1238–1245.

Oppenheimer J.R. & Snyder H. (1939), On Continued Gravitational Contraction. *Phys. Rev.* **56**: 455–459.

Oppenheimer J.R. & Volkoff G. (1939), On Massive Neutron Cores. *Phys. Rev.* **55**: 374–381.

Pak N. & Tze H.C. (1979), Chiral Solutions and Current Algebra. *Ann. Phys.* **117**: 164–194.

Papapetrou A. (1945), A Static Solution of the Gravitational Field for an Arbitrary Charge–Distribution. *Proc. Roy. Irish Acad.* **51**: 191–204.

Papapetrou A. (1953), Eine Rotationssymmetrische Lösung in der Allgemeinen Relativitätstheorie. *Ann. Physik.* **12**: 309–315.

Papapetrou A. (1966), Champs Gravitationels Stationaires a Symmetrie Axiale. *Ann. Inst. H. Poincaré A* **4**: 83–105.

Parker L. (1973), Conformal Energy–Momentum Tensor in Riemannian Space–Time. *Phys. Rev. D* **7**: 976–983.

Parker T. & Taubes C.H. (1982), On Witten's Proof of the Positive Energy Theorem. *Commun. Math. Phys.* **84**: 223–238.

Penrose R. (1963), Asymptotic Properties of Fields and Space–Times. *Phys. Rev. Lett.* **10**: 66–70.

Penrose R. (1964), Conformal Treatment of Infinity. In: *Relativity, Groups and Topology; The 1963 Les Houches Lectures,* eds. B. DeWitt & C. DeWitt. New York, Gordon and Breach.

Penrose R. (1965a), Zero Rest Mass Fields Including Gravitation: Asymptotic Behavior. *Proc. R. Soc. London Ser. A* **284**: 159–203.

Penrose R. (1965b), Gravitational Collapse and Space–Time Singularities. *Phys. Rev. Lett.* **14**: 57–59.

Penrose R. (1969), Gravitational Collapse: The Role of General Relativity. *Riv. Nuovo Cimento* **1**: 252–276.

Penrose R. (1973), Naked Singularities. *Ann. N.Y. Acad. Sci.* **224**: 125–134.

Penrose R. (1979), Singularities and Time Asymmetry. In: *General Relativity, an Einstein Centenary Survey,* eds. S.W. Hawking & W. Israel. Cambridge Univ. Press.

Perjes Z. (1971), Solutions of the Coupled Einstein–Maxwell Equations Representing the Fields of Spinning Sources. *Phys. Rev. Lett.* **27**: 1668–1670.

Rácz I. (1993), Maxwell Fields in Spacetimes Admitting Non–Null Killing Vectors. *Class. Quantum Grav.* **10**: L164–L172.

Rácz I. & Wald R. M. (1992), Extension of Spacetimes with Killing Horizons. *Class. Quantum Grav.* **9**: 2643–2656.

Rácz I. & Wald R. M. (1995), Global Extensions of Spacetimes Describing Asymptotic Finite States of Black Holes. gr-qc/9507055; to appear in *Class. Quantum Grav.*

Raychaudhuri A. (1955) Relativistic Cosmology. I. *Phys. Rev.* **98**: 1123–1126.

Reissner H. (1916), Ueber die Eigengravitation des Elektrischen Feldes nach der Einsteinschen Theorie. *Ann. Physik.* **50**: 106–120.

Robinson D.C. (1974), Classification of Black Holes with Electromagnetic Fields. *Phys. Rev.* **10**: 458–460.

Robinson D.C. (1975), Uniqueness of the Kerr Black Hole. *Phys. Rev. Lett.* **34**: 905–906.

Robinson D.C. (1977), A Simple Proof of the Generalization of Israel's Theorem. *Gen. Rel. Grav.* **8**: 695–698.

Ruback P. (1988), A New Uniqueness Theorem for Charged Black Holes. *Class. Quantum Grav.* **5**: L155–L159.

Sachs R.K. (1962), Asymptotic Symmetries in Gravitational Theory. *Phys. Rev.* **128**: 2851–2846.

Sachs R.K. (1964), Gravitational Radiation. In: *Relativity, Groups and Topology; The 1963 Les Houches Lectures*, eds. B. DeWitt & C. DeWitt. New York, Gordon and Breach.

Sacks G. & Uhlenbeck K. (1981), The Existence of Minimal Immersions of Two–Spheres. *Ann. Math.* **113**: 1–24.

Schmidt B. (1978), Null Infinity and Killing Fields. *J. Math. Phys.* **21**: 862–867.

Schoen R. & Uhlenbeck K. (1982), A Regularity Theory for Harmonic Maps. *J. Diff. Geom.* **17**: 307–335.

Schoen R. & Uhlenbeck K. (1983), Boundary Regularity and the Dirichlet Problem for Harmonic Maps. *J. Diff. Geom.* **18**: 253–268.

Schoen R. & Yau S.-T. (1979), On the Proof of the Positive Mass Conjecture in General Relativity. *Commun. Math. Phys.* **65**: 45–76.

Schoen R. & Yau S.-T. (1981), Proof of the Positive Mass Theorem. *Commun. Math. Phys.* **79**: 231–260.

Schwarzschild K. (1916a), Ueber das Gravitationsfeld eines Massenpunktes nach der Einsteinschen Theorie. *Sitz.ber. Deut. Akad. Wiss. Berlin, Kl. Math.-Phys. Tech.* 189–196.

Schwarzschild K. (1916b), Ueber das Gravitationsfeld einer Kugel aus inkompressibler Flüssigkeit nach der Einsteinschen Theorie. *Sitz.ber. Deut. Akad. Wiss. Berlin, Kl. Math.-Phys. Tech.* 424–434.

Simon W. (1984), The Multipole Expansion of Stationary Einstein–Maxwell Fields. *J. Math. Phys.* **25**: 1035–1038.

Simon W. (1985), A Simple Proof of the Generalized Israel Theorem. *Gen. Rel. Grav.* **17**: 761–768.

Simon W. (1992), Radiative Einstein–Maxwell Spacetimes and "No–Hair" Theorems. *Class. Quantum Grav.* **9**: 241–256.

Simon W. & Beig R. (1983), The Multipole Structure of Stationary Space–Times. *J. Math. Phys.* **24**: 1163–1171.

Skyrme T.H.R. (1961), A Non–Linear Field Theory. *Proc. R. Soc. London Ser. A* **260**: 127–138.

Skyrme T.H.R. (1962), A Unified Theory of Mesons and Baryons. *Nucl. Phys.* **31**: 556–559.

Skyrme T.H.R. (1971), Kinks and the Dirac Equation. *J. Math. Phys.* **12**: 1735–1743.

Smarr L. (1973), Mass Formula for Kerr Black Holes. *Phys. Rev. Lett.* **30**: 71–73. (erratum **30**: 521.)

Spivak M. (1979), *A Comprehensive Introduction to Differential Geometry*. Berkley, Publish or Perish Inc.

Straumann N. (1984), *General Relativity and Relativistic Astrophysics*. Berlin, Springer.

Straumann N. (1992), *Fields along Mappings between Manifolds*. Lecture Notes, unpublished.

Sudarsky D. (1995), A Simple Proof of a No Hair Theorem in Einstein Higgs Theory. *Class. Quantum Grav.* **12**: 579–584.

Sudarsky D. & Wald R.M. (1992), Extrema of Mass, Stationarity and Staticity, and Solutions to the Einstein–Yang–Mills Equations. *Phys. Rev. D* **46**: 1453–1474.

Sudarsky D. & Wald R.M. (1993), Mass Formulas for Stationary Einstein–Yang–Mills Black Holes and a Simple Proof of Two Staticity Theorems. *Phys. Rev. D* **47**: R5209–R5213.

Taubes C.H. & Parker T. (1982), On Witten's Proof of the Positive Energy Theorem. *Commun. Math. Phys.* **84**: 223–238.

Tod K.P. (1983), All Metrics Admitting Supercovariantly Constant Spinors. *Phys. Lett. B* **121**: 241–244.

Tomimatsu A. & Sato H. (1972), New Exact Solution for the Gravitational Field of a Spinning Mass. *Phys. Rev. Lett.* **29**: 1344–1345.

Tomimatsu A. & Sato H. (1973), New Series of Exact Solutions for Gravitational Fields of Spinning Masses. *Progr. Theor. Phys. (Kyoto)* **50**: 95–110.

Uhlenbeck K. (1989), Harmonic Maps into Lie Groups (Classical Solutions of the Chiral Model). *J. Diff. Geom.* **30**: 1–50.

Vishveshwara C.V. (1968), Generalization of the "Schwarzschild Surface" to Arbitrary Static and Stationary Metrics. *J. Math. Phys.* **9**: 1319–1322.

Visser M. (1992), Dirty Black Holes: Thermodynamics and Horizon Structure. *Phys. Rev. D* **46**: 2445–2451.

Volkov M.S. & Gal'tsov D.V. (1989), Non–Abelian Einstein–Yang–Mills Black Holes. *JETP Lett.* **50**: 346–350.

Wald R.M. (1974), Gedankenexperiments to Destroy a Black Hole. *Ann. Phys.* **82**: 548–556.

Wald R.M. (1975), On Particle Creation by Black Holes. *Commun. Math. Phys.* **45**: 9–34.

Wald R.M. (1984), *General Relativity.* Univ. of Chicago Press.

Wald R.M. (1992), Black Holes and Thermodynamics. In: *Black Hole Physics*, eds. V. de Sabbata & Z. Zhang. Series C: Math. and Phys. Sciences, Vol 364, The Netherlands, Kluwer Academic Publishers.

Wald R.M. (1993a), The First Law of Black Hole Mechanics. In: *Directions in General Relativity, Vol 1*, eds. B.L. Hu, M.P. Ryan & C.V. Vishveshwara. Cambridge Univ. Press.

Wald R.M. (1993b), Black Hole Entropy is the Noether Charge. *Phys. Rev. D* **48**: R3427–R3431.

Wald R.M. (1994), *Quantum Field Theory in Curved Spacetime and Black Hole Thermodynamics.* Univ. of Chicago Press.

Weinstein G. (1990), On Rotating Black Holes in Equilibrium in General Relativity. *Commun. Pure Appl. Math.* **43**: 903–948.

Weinstein G. (1992), The Stationary Axisymmetric Two-Body Problem in General Relativity. *Commun. Pure Appl. Math.* **45**: 1183–1203.

Weinstein G. (1994a), On the Force between Rotating Co-Axial Black Holes. *Trans. Am. Math. Soc.* **343** No. 2: 899–906.

Weinstein G. (1994b), N-Black Hole Stationary and Axially Symmetric solutions of the Einstein–Maxwell Equations gr-qc/9412036.

Westenholz G., von (1978), *Differential Forms in Mathematical Physics.* New York, North–Holland Publishing.

Weyl H. (1917), Zur Gravitationstheorie. *Ann. Physik. (Leipzig, 4.f.)* **54**: 117–145.

Weyl H. (1919), Bemerkung über die axisymmetrischen Lösungen der Einsteinschen Gravitationsgleichungen. *Ann. Physik. (Leipzig, 4.f.)* **59**: 185–188.

Wheeler J.A. (1968), Our Universe: The Known and the Unknown. *American Scientist* **56**: 1–20.

Willmore T.Y. (1993), *Riemannian Geometry.* Oxford, Clarendon Press.

Witten E. (1981), A New Proof of the Positive Energy Theorem. *Commun. Math. Phys.* **80**: 381–402.

Witten E. (1983), Global Aspects of Current Algebra. *Nucl. Phys. B* **223**: 422–444.

Xanthopoulos B.C. (1978), Isometries Compatible with Asymptotic Flatness at Null Infinity. *J. Math. Phys.* **19**: 2216–2222.

Xanthopoulos B.C. & Zannias T. (1991), The Uniqueness of the Bekenstein Black Hole. *J. Math. Phys.* **32**: 1875–1880.

Zahed I. & Brown G. E. (1986), The Skyrme Model. *Phys. Rep.* **142**: 1–102.

Index

Printed in the United States
By Bookmasters